Burhan Ahmad
Shahid Mahmood

Observed, Simulated and Projected Extreme Climate Indices over Pakistan

Anchor Academic
Publishing

Ahmad, Burhan, Mahmood, Shahid: Observed, Simulated and Projected Extreme Climate Indices over Pakistan, Hamburg, Anchor Academic Publishing 2017

Buch-ISBN: 978-3-96067-172-5
PDF-eBook-ISBN: 978-3-96067-672-0
Druck/Herstellung: Anchor Academic Publishing, Hamburg, 2017

Bibliografische Information der Deutschen Nationalbibliothek:
Die Deutsche Nationalbibliothek verzeichnet diese Publikation in der Deutschen Nationalbibliografie; detaillierte bibliografische Daten sind im Internet über http://dnb.d-nb.de abrufbar.

Bibliographical Information of the German National Library:
The German National Library lists this publication in the German National Bibliography. Detailed bibliographic data can be found at: http://dnb.d-nb.de

All rights reserved. This publication may not be reproduced, stored in a retrieval system or transmitted, in any form or by any means, electronic, mechanical, photocopying, recording or otherwise, without the prior permission of the publishers.

Das Werk einschließlich aller seiner Teile ist urheberrechtlich geschützt. Jede Verwertung außerhalb der Grenzen des Urheberrechtsgesetzes ist ohne Zustimmung des Verlages unzulässig und strafbar. Dies gilt insbesondere für Vervielfältigungen, Übersetzungen, Mikroverfilmungen und die Einspeicherung und Bearbeitung in elektronischen Systemen.

Die Wiedergabe von Gebrauchsnamen, Handelsnamen, Warenbezeichnungen usw. in diesem Werk berechtigt auch ohne besondere Kennzeichnung nicht zu der Annahme, dass solche Namen im Sinne der Warenzeichen- und Markenschutz-Gesetzgebung als frei zu betrachten wären und daher von jedermann benutzt werden dürften.

Die Informationen in diesem Werk wurden mit Sorgfalt erarbeitet. Dennoch können Fehler nicht vollständig ausgeschlossen werden und die Diplomica Verlag GmbH, die Autoren oder Übersetzer übernehmen keine juristische Verantwortung oder irgendeine Haftung für evtl. verbliebene fehlerhafte Angaben und deren Folgen.

Alle Rechte vorbehalten

© Anchor Academic Publishing, Imprint der Diplomica Verlag GmbH
Hermannstal 119k, 22119 Hamburg
http://www.diplomica-verlag.de, Hamburg 2017
Printed in Germany

Executive Summary

This research explores observed, simulated, and projected extreme climate indices from a selection of different GCMs from CMIP5 ensemble for Pakistan at province level. The extreme indices for observed, simulated and projected climate are found and analysed on provincial basis over the country.

Results show that summer days in the GB-AJK are increasing significantly (by 24 days in Gilgit and by 18 days in Muzaffarabad) which may trigger high frequencies of Glacier Lake Outburst Floods in those regions. Projected climate has suggested an added increase of 6 days in the summer days index over the GB-AJK region in the next 30 years. Significant increasing trends in very heavy precipitation extremes are seen in the KPK province which may bring flood related liabilities in the region. Peshawar has shown a 0.9°C increase in the past mean maximum temperature index, whereas, the KPK province has further been projected to show a whopping 1.3°C added increase in the mean maximum temperature by the year 2045. This work has seen high confidence in the trend increase of minimum temperature related indices in the Punjab province, which implicates agricultural hazards to the food security of the country. Average increase in the mean minimum temperatures is 1.9°C over the past 54 years over the Punjab province. Projected index for the mean minimum temperature in Punjab has suggested an added 1.4°C increase in the next 30 years. The mean maximum temperatures, the summer days, and the warm days, all display significant increasing trends in the Sindh province, which may evolve high frequency and prolonged heat waves in the region. Number of heat waves days in Karachi have increased by 8 days over the historical period, while they are suggested to increase further by 20 days in the next 30 years projection period over the Sindh province. Also, significant increasing trends in temperature related indices are liable to bring thermal instability in the Balochistan province, which ultimately brings the potential of high frequency dust storms over the region.

Pakistan has been facing shortages in both power and water sector which are lifelines of a country. Significant increases in the maximum and minimum temperatures over the country may affect such sectors drastically. Considerable increase in the frequency and intensity of extreme weather events, coupled with erratic monsoon rains may cause frequent and intense floods and droughts in the region. Rising temperatures resulting in enhanced heat and water-stressed conditions, particularly in arid and semi-arid regions, may lead to reduced agricultural productivity. This report shall bring added value to all the stake holders and the policy maker in determining the hazards that extreme climate has brought in the past and may bring in the near future.

Table of Contents

1 Introduction 1
   1.1 Description of Regional Climate in Pakistan 1
   1.2 Climate Induced Natural Disasters in Pakistan 2
   1.3 Literature Review 4

2 Data and Methodology 7
   2.1 Data 7
      2.1.1 PMD observed data 7
      2.1.2 GCM simulated data 8
      2.1.3 High resolution statistically downscaled GCM data 8
      2.1.4 High resolution dynamically downscaled GCM data 10
   2.2 Methodology 10

3 Verification of GCMs and RCMs 13
   3.1 GCMs emulating observed climatology 13
      3.1.1 GCMs emulating observed climatology over Baluchistan 13
      3.1.2 GCMs emulating observed climatology over GB-AJK 14
      3.1.3 GCMs emulating observed climatology over KPK 14
      3.1.4 GCMs emulating observed climatology over Punjab 15
      3.1.5 GCMs emulating observed climatology over Sindh 15
   3.2 Statistically downscaled GCMs emulating observed climatology 18
      3.2.1 Statistically downscaled GCMs emulating observed climatology over KPK 18
      3.2.2 Statistically downscaled GCMs emulating observed climatology over Punjab 18
      3.2.3 Statistically downscaled GCMs emulating observed climatology over Sindh 19
      3.2.4 Statistically downscaled GCMs emulating observed climatology over Baluchistan 19
      3.2.5 Statistically downscaled GCMs emulating observed climatology over GB-AJK 19
   3.3 Dynamically downscaled GCMs emulating observed climatology 22
      3.3.1 Dynamically downscaled GCMs emulating observed climatology over GB-AJK 22
      3.3.2 Dynamically downscaled GCMs emulating observed climatology over KPK 23

    3.3.3   Dynamically downscaled GCMs emulating observed climatology over Punjab 23

    3.3.4   Dynamically downscaled GCMs emulating observed climatology over Sindh 24

    3.3.5   Dynamically downscaled GCMs emulating observed climatology over Baluchistan 24

  3.4   Verdict over GCMs and RCMs subsets selection 27

4   Types of climate indices 31

  4.1   Percentile-based indices 31

  4.2   Absolute indices 31

  4.3   Threshold indices 31

  4.4   Duration indices 31

  4.5   Societal Impact indices 32

5   Results and discussion of observed climate indices 33

  5.1   Observed indices of GB-AJK 33

  5.2   Observed indices of KPK 35

  5.3   Observed indices of Punjab 38

  5.4   Observed indices of Sindh...40

  5.5   Observed indices of Baluchistan 44

  5.6   Observed indices of All Pakistan 54

6   GCMs and RCMs simulated indices 57

  6.1   GCM ensembles simulated indices 57

    6.1.1   GCM ensembles simulated indices over GB-AJK 57

    6.1.2   GCM ensembles simulated indices over KPK 57

    6.1.3   GCM ensembles simulated indices over Punjab 58

    6.1.4   GCM ensembles simulated indices over Sindh..58

    6.1.5   GCM ensembles simulated indices over Baluchistan 59

  6.2   Statistically downscaled GCMs (Nex-NASA) based climate indices...............59

    6.2.1   Nex-NASA based climate indices over GB-AJK 59

    6.2.2   Nex-NASA based climate indices over KPK..60

    6.2.3   Nex-NASA based climate indices over Punjab 60

    6.2.4   Nex-NASA based climate indices over Sindh...61

    6.2.5   Nex-NASA based climate indices over Baluchistan 61

  6.3   Dynamically downscaled (CORDEX) based climate indices...........................61

    6.3.1   CORDEX based climate indices over GB-AJK 62

    6.3.2   CORDEX based climate indices over KPK...62

	6.3.3	CORDEX based climate indices over Punjab	63
	6.3.4	CORDEX based climate indices over Sindh	64
	6.3.5	CORDEX based climate indices over Baluchistan	65
7		Projected Climate and Indices over the provinces of Pakistan	67
	7.1	Next 30 year's ensemble projected indices	67
	7.1.1	Projected indices over Punjab	67
	7.1.2	Projected indices over Baluchistan	68
	7.1.3	Projected indices over GB-AJK	70
	7.1.4	Projected indices over KPK	72
	7.1.5	Projected indices over Sindh	73
	7.2	Projected climate over Pakistan by 2050	75
	7.2.1	Projected climate in GB-AJK	75
	7.2.2	Projected climate in KPK	75
	7.2.3	Projected climate in Punjab	76
	7.2.4	Projected climate over Sindh	77
	7.2.5	Projected climate in Baluchistan	77
	7.2.6	Projected climate over All Pakistan	78
8		Discussion	79
9		Conclusions	79
	9.1	GB-AJK	79
	9.2	KPK	80
	9.3	Punjab	80
	9.4	Sindh	80
	9.5	Baluchistan	81
10		References	83
11		Appendix 1	87
	11.1	Observed climate indices (1960-2013)	87
	11.2	Ensemble mean GCMs simulated climate indices (1970-2004)	87
	11.3	Nex-NASA based climate indices (1970-2004)	87
	11.4	CORDEX based climate indices (1970-2004)	87
	11.5	Nex-NASA based projected climate indices (2016-2045)	88

LIST OF FIGURES

Figure 1.1: Spatial distribution of normal climate of Pakistan (1960-2013) 2
Figure 2.1: Cressman interpolated maximum temperature, minimum temperature and precipitation normal climate (1960-2013) over Pakistan. 7
Figure 2.2: Station-wise missing climate data percentage shown both graphically and spatially. 11
Figure 2.3: Province-wise selected stations with relief in meters. 12
Figure 3.1: Observed and GCMs annual cycles of TMAX (°C), TMIN (°C) and Precip (mm) over 5 different provinces for 1970-2004 historical period. 16
Figure 3.2: Province based Taylor Diagrams for TMAX, TMIN, and Precip, for the raw GCMs used in this study. 17
Figure 3.3: Observed and statistically downscaled Nex-NASA GCMs annual cycles of TMAX (°C), TMIN (°C) and Precip (mm) over 5 different provinces for 1970-2004 historical period. 20
Figure 3.4: Province based Taylor Diagrams for TMAX, TMIN, and Precip, for the Nex-NASA GCMs used in this study. 21
Figure 3.5: Observed and dynamically downscaled CORDEX GCMs annual cycles of TMAX (°C), TMIN (°C) and Precip (mm) over 5 different provinces for 1970-2004 historical period. 25
Figure 3.6: Province based Taylor Diagrams for TMAX, TMIN, and Precip, for the CORDEX RCMs used in this study. 26
Figure 3.7: GCM grids and province-wise grids count for selected GCMs. 28
Figure 5.1: Significant observed climate indices over GB-AJK region. 47
Figure 5.2: Significant observed climate indices over KPK province. 48
Figure 5.3: Significant observed climate indices over Punjab province. 49
Figure 5.4a: Significant observed climate indices over Sindh province 50
Figure 5.4b: Significant observed climate indices over Sindh province (Continued) 51
Figure 5.5a: Significant observed climate indices over Baluchistan province. 52
Figure 5.5b: Significant observed climate indices over Baluchistan province (continued). 53
Figure 5.6: Significant observed climate indices over whole Pakistan. 55
Figure 5.7: Spatial historical (1960-2013) climate trends over Pakistan 56
Figure 6.1: Significant simulated, statistically downscaled, and dynamically downscaled climate indices over GB-AJK region. 62
Figure 6.2: Significant simulated, statistically downscaled, and dynamically downscaled climate indices over KPK province. 63
Figure 6.3: Significant simulated, statistically downscaled, and dynamically downscaled climate indices over Punjab province. 64
Figure 6.4: Significant simulated, statistically downscaled, and dynamically downscaled climate indices over Sindh province. 65
Figure 6.5: Significant simulated, statistically downscaled, and dynamically downscaled climate indices over Baluchistan province. 66
Figure 7.1: Next 30 year's (2016-2045) ensemble projected indices over Punjab province. 68
Figure 7.2: Next 30 year's (2016-2045) ensemble projected indices over Baluchistan province. 70
Figure 7.3: Next 30 year's (2016-2045) ensemble projected indices over GB-AJK region. 71
Figure 7.4: Next 30 year's (2016-2045) ensemble projected indices over KPK province. 73
Figure 7.5: Next 30 year's (2016-2045) ensemble projected indices over Sindh province. 74

List of Tables

Table 2.1: PMD stations with coordinates and elevations...9
Table 2.2: GCMs with their modeling institutes and horizontal resolutions......................................10
Table 3.1: Summary of the selected raw GCMs, the statistically downscaled GCMs, and the dynamically downscaled GCMs. 29
Table 4.1: Core climate indices calculated and analyzed in this study. 32
Table 7.1: Historical and projected seasonal climate over Pakistan 78

1 INTRODUCTION

1.1 DESCRIPTION OF REGIONAL CLIMATE IN PAKISTAN

Pakistan is situated between 23°N to 37°N and 60°E to 76°E, on the western confines of the SW Monsoon (SWM) over subcontinent. Its climate is mainly arid to semiarid having a small area with sub humid climate. Pakistan being an agricultural state depends on water resources like rainfall, glacial ice melt etc. Precipitation is received in both summer and winter seasons. Western disturbances approach the upper parts of the country throughout the year while Easterlies are dominant in summer. In summer heavy falls are mainly due to superposition of easterly and westerly.

The dominant air mass that impacts Pakistan is mostly Tropical (due to low latitude and warm weather) and mostly Continental (although one part of it is in contact with the ocean, the land is mostly inland). Towards the coast, the climate varies from warm to breezy, however as the country progresses more inland, there are higher elevations and generally cooler temperatures. The changing of the seasons is characterized by a cold, dry winter, a hot and dry winter, and a wet summer of monsoons. Pakistan is affected by the trade winds. Pakistan is also in close proximity to the Indian Ocean Gyre. The most abundant source of rainfall in Pakistan is in the form of the southwest monsoon, a natural occurrence that causes intense floods. It is common because of Pakistan's proximity to the cold regions of Asia (Himalayas). Monsoons blow from the cold to warm regions so the wind that blow through regions such as the Himalayas bring the Monsoons to Pakistan.

Pakistan is located between the upper-level easterly subtropical jet to the south and the westerly mid-latitude jet to the north. At mid-levels, typically dry and diurnally heated air flows off the Afghan Plateau. This dry subsiding flow occasionally provides a stable cap over low-level moisture advected into the foothills of northern Pakistan from the Arabian Sea, and intense deep convection in the northwest indentation of the Himalayas occurs when convection breaks through the cap. Mean conditions at both 700 and 850 hPa indicate low-to mid-level westerly wind components throughout south to central India. During the monsoon season, atmospheric moisture is maximized over the Bay of Bengal and Arabian Sea as evaporative latent heat fluxes act to increase the moisture content in the boundary layer. This moist air flows onto the continent in a generally southwesterly direction in response to the thermally driven monsoon circulation. As the moist air encounters topographic features of various scales, in the Hindu Kush and Sulaiman ranges of Afghanistan, the Himalayas, and the Arakan range of Burma, precipitation tends to be maximized upstream of and over the windward slopes of such features. However, the typical monsoon precipitation in Pakistan is notably small with the exception of the northern part of the country that experiences frequent deep convection near the intersection of the Hindu Kush and Himalayan ranges. Pakistan is generally dry and arid, even during the monsoon season; thus, anomalously large amounts of precipitation in the region are unexpected and can have catastrophic impacts (for normal climate of Pakistan, see Figure 1.1).

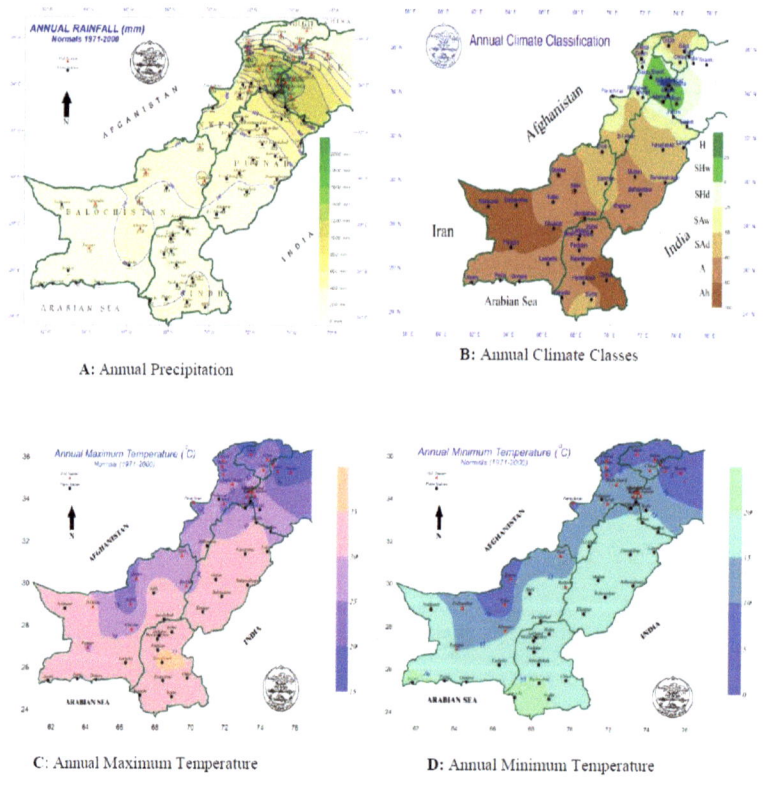

Figure 1.1: Spatial distribution of normal climate of Pakistan (1960-2013) (Source: PMD 2009)

1.2 Climate Induced Natural Disasters in Pakistan

A mini TORNADO (Twister) had hit Peshawar valley with a speed of 110 km/hr and reported rainfall of 60 mm on 27th April, 2015, causing huge damages. It was very unusual extreme weather event that rarely occur in Pakistan. In the past, same type of windstorms (Tornados) occurred as;

1. In March 2001, in Village "Chak Misran", Tehsil Bhalwal, District Sargodha with a speed of 193 km/hr, 20 reported dead & 40 injured.

2. In March 2011, in Village "Bahadur Pur",near Head Marala, District Sialkot with a speed of 100 km/hr, 10 reported dead & 32 injured.

According to the climate of Pakistan, such types of extreme weather events rarely develop due to thermal contrast during the months of March or April (a transition period, when days are getting hot and nights are still cold) due to interaction of cold air mass (from westward) & warm air mas (from east or south). For Peshawar, "Rain with Dust-Thunderstorm" was predicted two days before by PMD but the prediction of tornado or twister is not possible in the world, even today's advancement in radar and satellite technologies. The occurrence of such high impact weather events (extreme) revealed that the consequences of Climate Change, due to deforestation, urbanization & pollution have already happening in Pakistan and the probability of such extreme events would be high in the future.

People of Pakistan witnessed a heatwave event in southern parts of the country including from 17th to 24th June 2015. This heatwave event took more than 1200 human lives in Karachi and about 200 lives in other parts of Sindh province. Main cause of deaths was heatstroke and dehyderation during extreme temperatures. In Karachi maximum temperature of 44.8°C was recorded which is second highest temperature after the year 1979. The intensity of heat index or apparent temperature was calculated to be 66°C on the peak heatwave day i.e. 20th June 2015. The heat wave was a symptom of global climate change, aggravated by deforestation, expansion of asphalt superhighways, and rapid urbanisation.

On the night falling between 15th and 16th July, and again on 19th, 24th and 28thJuly 2015, different parts of District Chitral were hit by Glacial Lake Outburst Floods (GLOF) which carried massive torrents and flash floods washing away villages, access roads, bridges, drinking water supply systems, micro hydel power channels, public/private property and agricultural crops. This phenomenon is unusual; however, due to climate changes, districts in the north of Khyber Pakhtunkhwa i.e. Mansehra, Chitral,

Battagram, Kohistan (U/L), Torghar and Shangla are prone to such incidents of GLOF. Met Office predicted that monsoon currents would penetrate in upper parts of the country in coming days. A westerly wave was also likely to approach northern areas of the country during that weekend. Scattered rain/thundershowers were expected in Punjab, Khyber Pakhtunkhwa, FATA, Gilgit Baltistan and Kashmir during Saturday to Monday with isolated heavy falls in Malakand, Hazara, Mardan, Peshawar, Rawalpindi, Gujranwala and Lahore divisions, Upper FATA and Kashmir. The clear sky conditions and high temperatures of northern areas during next two days had a potential to produce more glacier-melt during the weekend. Rainfall over the glaciers triggered the melt rate, causing more flash floods in the local rivers and streams of the northern areas (Gilgit-Baltistan and Malakand Division including Chitral) during that weekend. There was an increased risk of Glacial Lake Outburst Flood (GLOF) in Gilgit-Baltistan and District Chitral during the period. Local communities were advised to remain alert and concerned authorities to take precautionary measures.

The earliest snowfall in 40 years had dumped up to four feet of snow on one town in less than 24 hours. Snowstorm in Babusar area of Pakistan had claimed life of a person. Snowfall came early this year. It started lashing Kaghan Valley on 24th Oct 2015 and continued intermittently for a second consecutive day on 25th Oct 2015. The area from Naran to Babusar Top received up to four feet of snow, leaving the roads blocked and over 1,200 tourists, including women and children, from across the country stranded. They faced difficulties as they ran out of supplies amidst continuous rain and snowfall. The snowfall has also damaged the power lines and some small bridges. Met office predicted snowfall over the hills of Malakand division, GB and Kashmir during the period.

1.3 Literature Review

The anthropogenic climate change and its impacts are worrying the human race. The last century has seen significant increase in global surface temperatures and, in addition, this process of increasing global temperatures will continue (Houghton, et al., 2001), as mentioned in IPCC report, until the emission of CFCs are not controlled. It is important to realize that the need and limited human resources and technology drive towards generation of projections by sampling GCMs and RCP scenarios that are practicable to cultivate, examine and publicize. In addition, these projections should be policy relevant for the region (McSweeney, et al., 2015).

Coupled Model Intercomparision Project Phase 5 (CMIP5), as compared to CMIP3, aggregates superior global climate models (Sillmann, et al., 2013). In addition, the availability of the simulations, generated from modern climate models (GCM), encourage climate researchers do their climate research within the CMIP5. This research can include not only mean climate but the weather-extremes as well. As the extremes occur once a blue moon, therefore, to mark its period and strength, it is necessary to explore long duration data sets of actual data (Houghton, et al., 2001). In the CMIP5 experiment it was agreed to grant access to the instantaneous fields of prognostic variables from GCMs to be used as lateral boundary conditions (LBCs) to steer regional climate models (RCMs). It opened the gates for researchers to downscale with diverse combinations of GCMs and RCMs or statistical downscaling methods to explore variety of high-resolution projections for different regional areas of the world (McSweeney, et al., 2015).

It is accepted by all that environment and humanities are effected severely with the changes in frequency and intensity of extreme events (Sohrabi, et al., 2013). Indices based on Percentile, threshold, duration and on other statistical tools can be used to describe extreme events (Houghton, et al., 2001). To examine the extremes, the Expert Team on Climate Change Detection and Indices (ETCCDI) has introduced a set of climate indices which are used globally for analyzing changing extremes in observed data (Sillmann, et al., 2013). These indices are also used for future climate projections. The ETCCDI indices help in conceiving a detailed picture of temperature and precipitation over a region (Sillmann, et al., 2013). The study of extreme events require data from greater number of long term weather stations of the region. But, lack of funding, human resource, technology and political instability are the main obstacles in developing countries in maintaining healthy and quality controlled weather data repositories (Choi, et al., 2009). According to (Sillmann, et al., 2008) a number of researchers focused on data gathered from ground weather stations and meteorological observatories as (Frich, et al., 2002) (Klein Tank, et al., 2003) did for studying climatic extreme indices while several also used future climate projections like (Meehl, et al., 2000) (Meehl, et al., 2004) (Tebaldi, et al., 2006).

The key to build mitigation and adaptation plans against extreme events due to climate change, microscopic examination of extreme events in terms of intensity, duration and frequency is necessary (Choi, et al., 2009). An effort is made to comprehensively evaluate extreme indices on the scale of modern CMIP5 models for Pakistan, at province level. This type of comparison between indices calculated from observational data and multi-model simulations has not been done before, specially keeping the policy makers for provinces in mind. It has been seen that multimodel ensemble simulations are quite better than individual models in providing healthier estimates of future climate variations and model related reservations (Sillmann, et al., 2013).

This report explores the selection of different GCMs from CMIP5 ensemble for Pakistan at province level. In this way we found 5 different subsets of CMIP5 GCMs for 5 different province groups. Moreover, the same technique is also adopted for CORDEX (McSweeney, et al., 2015) and Nex-NASA. Afterwards, extreme indices for observed, simulated and projected climate are found and analyzed on provincial basis over the country.

2 DATA AND METHODOLOGY

2.1 DATA

Four types of climatic datasets with 3 parameters each (maximum temperature, minimum temperature, and rainfall) over Pakistan are used in this study; 1. PMD observed data, 2. GCM simulated data, 3. High resolution statistically downscaled GCM data, and 4. High resolution dynamically downscaled GCM data.

2.1.1 PMD observed data

The PMD observed daily data is taken for the period 1960-2013 over a wide range of 50 stations spread all over the country. Details of the 50 stations are given in Table 2.1. Selection of stations includes elevation perspective as well as representation from all possible climatic zones of the country (Salma et al., 2012).

Cressman interpolation (see e.g., Goodin et al., 1979) is used to interpolate point data from 27 stations, each with 54 years' historical data length (1960-2013). The technique interpolates station data to a user-defined latitude-longitude grid. Multiple passes are made through the grid at consecutively smaller radii of influence to increase precision. Using the said technique, maximum temperature, minimum temperature, and precipitation are interpolated at a 0.5°×0.5° horizontal resolution. The three variables had also been interpolated previously with station data over a grid for a normal period (1970-2000) by Chaudhry et al., (2009). A spatial comparison of the two indicates good harmony among all the gridded variables interpolated previously, and those done under this study (Figure 2.1).

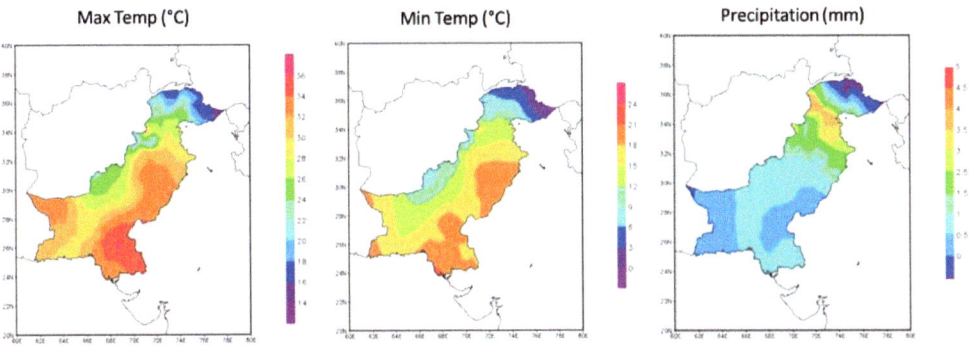

Figure 2.1: Cressman interpolated maximum temperature, minimum temperature and precipitation normal climate (1960-2013) over Pakistan.

2.1.2 GCM simulated data

GCMs daily simulated data for historical period (1970-2004) is taken under Coupled Model Intercomparison Project Phase 5 (CMIP5) protocols. Details of models are given in Table 2.2. GCMs describe three-dimensional geometry of the atmosphere and other components of Earth's climate system. Atmospheric GCMs numerically solve the equations of physics (e.g., dynamics, thermodynamics, radiative transfer, etc.) and chemistry applied to the atmosphere and its constituent components, including the greenhouse gases. The CMIP5 provides a multi-model context for assessing the mechanisms responsible for model differences in poorly understood feedbacks associated with the carbon cycle and with clouds. It also examines climate "predictability" and explores the ability of models to predict climate on decadal time scales, and, more generally, determines why similarly forced models produce a range of responses (Taylor et al., 2012).

2.1.3 High resolution statistically downscaled GCM data

Downscaling is a method for obtaining high-resolution climate or climate change information from relatively coarse-resolution global climate models (GCMs). Typically, GCMs have a resolution of 150-300 km by 150-300 km. Many impacts models require information at scales of 50 km or less, so some method is needed to estimate the smaller-scale information. Statistical downscaling first derives statistical relationships between observed small-scale (often station level) variables and larger (GCM) scale variables, using either analogue methods (circulation typing), regression analysis, or neural network methods. Future values of the largescale variables obtained from GCM projections of future climate are then used to derive the statistical relationships and so estimate the smaller-scale details of future climate (see e.g., Burhan et al., 2015a).

This study has used ensemble mean of three NEX Global Daily Downscaled Climate Models viz. "CanESM2" of Canadian Centre for Climate Modelling and Analysis, "CNRM-CM5" of Centre National de Recherches Meteorologiques, France, and "MRI-CGCM3" of Meteorological Research Institute, Japan, all bias corrected with quantile mapping (Thrasher et al., 2012). Common historical time period of 35 years (1970-2004) is taken for the analysis. The NASA Earth Exchange Global Daily Downscaled Projections (NEX-GDDP) dataset is comprised of downscaled climate scenarios for the globe that are derived from the GCM runs conducted under the CMIP5 and across two of the four greenhouse gas emissions scenarios known as Representative Concentration Pathways (RCPs).

Table 2.1:PMD stations with coordinates and elevations.

SR. NO	STATION	LAT (N)	LON (E)	ELV (m)	SR. NO	STATION	LAT (N)	LON (E)	ELV (m)
1	Gupis	36° 10'	73° 24'	2156	26	Lahore	31° 33'	74° 20'	216
2	Gilgit	35° 55'	74° 20'	1460	27	Faisalabad	31° 26'	73° 06'	185
3	Bunji	35° 40'	74° 38'	1372	28	Zhob	31° 21'	69° 28'	1405
4	Drosh	35° 34'	71° 47'	1463	29	Quetta	30° 15'	66° 53'	1719
5	Chilas	35° 25'	74° 06'	1250	30	Multan	30° 12'	71° 26'	121
6	Astore	35° 22'	74° 54'	2168	31	Bahawalnagar	29° 57'	73° 15'	161
7	Skardu	35° 18'	75° 41'	2317	32	Barkhan	29° 53'	69° 43'	1097
8	Dir	35° 12'	71° 51'	1375	33	Sibbi	29° 33'	67° 53'	133
9	Saidu Sharif	34° 44'	72° 21'	961	34	Bahawalpur	29° 24'	71° 47'	110
10	Balakot	34° 23'	73° 21'	995	35	Kalat	29° 02'	66° 35'	2015
11	Muzaffarabad	34° 22'	73° 29'	702	36	Dalbandin	28° 53'	64° 24'	848
12	Ghari Dupatta	34° 13'	73° 37'	813	37	Nokkundi	28° 49'	62° 45'	682
13	Kakul	34° 11'	73° 15'	1308	38	Khanpur	28° 39'	70° 41'	87
14	Peshawar	34° 01'	71° 35'	359	39	Jacobabad	28° 18'	68° 28'	55
15	Islamabad	33°42'	73°05'	508	40	Khuzdar	27° 50'	66° 38'	1231
16	Murree	33° 55'	73° 23'	2168	41	Rohri	27° 42'	68° 54'	66
17	Parachinar	33° 52'	70° 05'	1775	42	Panjgur	26° 58'	64° 06'	980
18	Cherat	33° 49'	71° 53'	1372	43	Padidan	26° 51'	68° 08'	46
19	Kohat	33° 34'	71° 26'	489	44	Nawabshah	26° 15'	68° 22'	37
20	Kotli	33° 31'	73° 54'	614	45	Chhor	25° 31'	69° 47'	5
21	Jhelum	32° 56'	73° 43'	234	46	Hyderabad	25° 23'	68° 25'	40
22	Mianwali	32° 33'	71° 33'	210	47	Pasni	25° 16'	63° 29'	4
23	Sialkot	32° 30'	74° 32'	255	48	Jiwani	25° 04'	61° 48'	56
24	Sargodha	32° 03'	72° 40'	187	49	Karachi	24° 54'	67° 08'	21
25	DIkhan	31° 49'	70° 55'	173	50	Badin	24° 38'	68° 54'	10

The CMIP5 GCM runs were developed in support of the Fifth Assessment Report of the Intergovernmental Panel on Climate Change (IPCC AR5). The NEX-GDDP dataset includes downscaled projections for RCP 4.5 and RCP 8.5 from 21 models and scenarios for which daily scenarios were produced and distributed under CMIP5. Each of the climate projections includes daily maximum temperature, minimum temperature, and precipitation for the periods from 1950 through 2100. The spatial resolution of the dataset is 0.25 degrees (~25 km x 25 km).

Table 2.2: GCMs with their modeling institutes and horizontal resolutions.

SR NO.	GCM	MODELLING CENTER	HORIZONTAL RESOLUTION
1	BCC-CSM1.1	Beijing Climate Center, China Meteorological Administration	2.8°
2	CanESM2	Canadian Centre for Climate Modelling and Analysis	2.8°
3	CCSM4	NCAR (National Center for Atmospheric Research) Boulder, CO, USA	1.25°
4	CSIRO-Mk3.6.0	Commonwealth Scientific and Industrial Research Organisation	1.8°
5	HadGEM2-AO	NIMR (National Institute of Meteorological Research, Seoul, South Korea)	1.8°
6	MIROC5	Atmosphere and Ocean Research Institute (The University of Tokyo), Japan	1.4°
7	MRI-CGCM3	Meteorological Research Institute, Japan	1.12°

2.1.4 High resolution dynamically downscaled GCM data

Dynamical downscaling uses a limited-area, high-resolution model (a regional climate model, or RCM) driven by boundary conditions from a GCM to derive smaller-scale information. RCMs generally have a domain area of 106 to 107 km2 and a resolution of 20 to 60 km. This study has used Coordinated Regional Climate Downscaling Experiment (CORDEX) data of six dynamically downscaled GCMs viz. CSIRO CCAM ACCESS-1, CSIRO CCAM CCSM4, CSIRO CCAM GFDL-CM3, CSIRO CCAM MPI-ESM-LR, REMO2009 MPI-ESM-LR, and SMHI RCA4 ICHEC-EC-EARTH, for South Asian region at 25 Km horizontal resolution. For the South Asian region, CORDEX presents an unprecedented opportunity to advance knowledge of regional climate responses to global climate change (see e.g., Burhan et al., 2015b), and for these insights to feed into Working Groups One and Two of the IPCC Fifth Assessment Report as well as to on-going climate adaptation and risk assessment research and policy planning in the region.

2.2 METHODOLOGY

Climate data from 50 PMD stations were first put to quality check and control using RClimDex (1.0) (Zhang et al., 2004). Missing years, months and days were sorted out and logged for each of the stations, and a missing percentage calculated for the selection of stations (Figure 2.1). It was found that years prior to 1970 had higher percentages of missing climate data. Moreover, higher percentages of missing data were concentrated to south-west regions of the country. The stations with higher percentage of continuously available climate data were selected for which the missing data percentage was less than 5% in 54 years. The selected stations (Figure 2.2) were then formatted for further quality check and control in RClimDex (1.0) software.

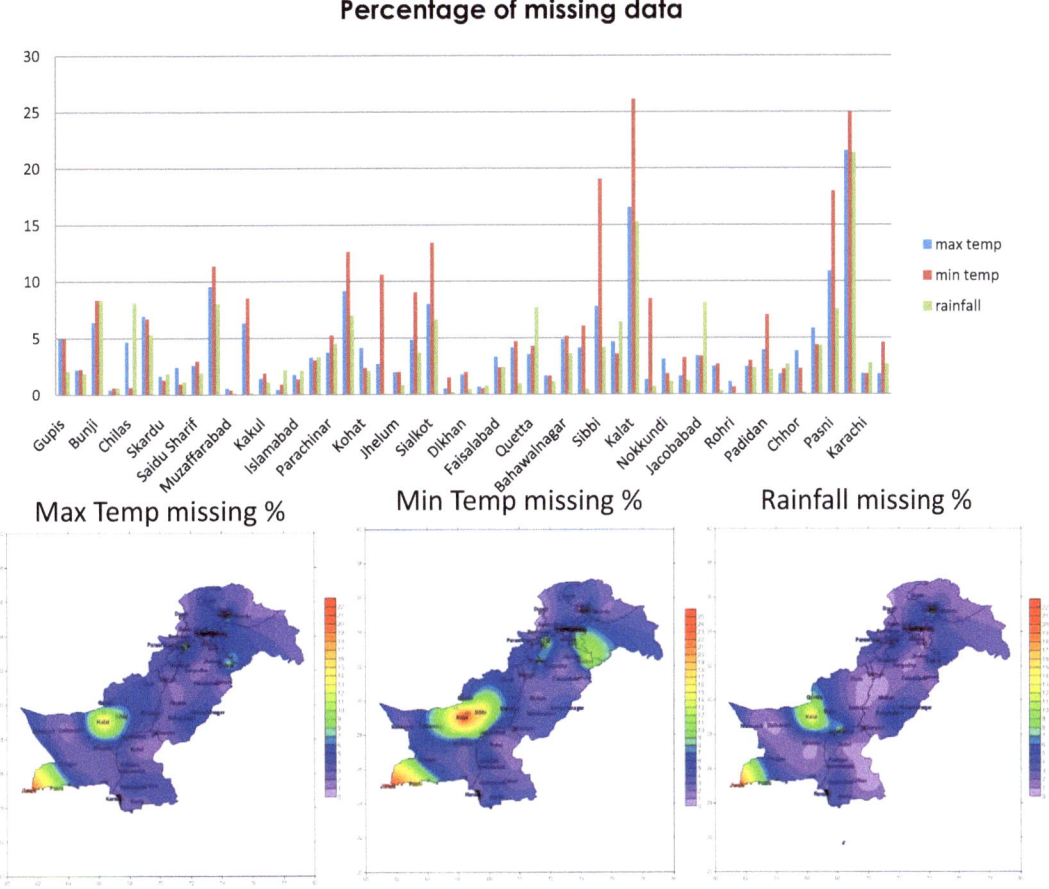

Figure 2.2: Station-wise missing climate data percentage shown both graphically and spatially.

The following checks were taken into consideration while quality controlling the data;

1. The outliers were identified and set to missing using a 3 × Standard Deviation criteria for maximum and minimum temperature.

2. For precipitation, corresponding daily values of monthly maxima were searched in "Monthly Climatic Normals of Pakistan", and set as upper limit thresholds for the identification of precipitation outliers.

3. Quality checks were also done for values showing minimum temperature ≥ maximum temperature.

After the quality check and control procedure, extreme climate indices were calculated with PMD observed data for 27 stations based on provincial division. Each selected station was strategically chosen so that it may deliver maximum coverage to the province it represents. For GCMs and RCMs, province-wise indices were calculated and compared with observed to find for identical climate extremes in the region.

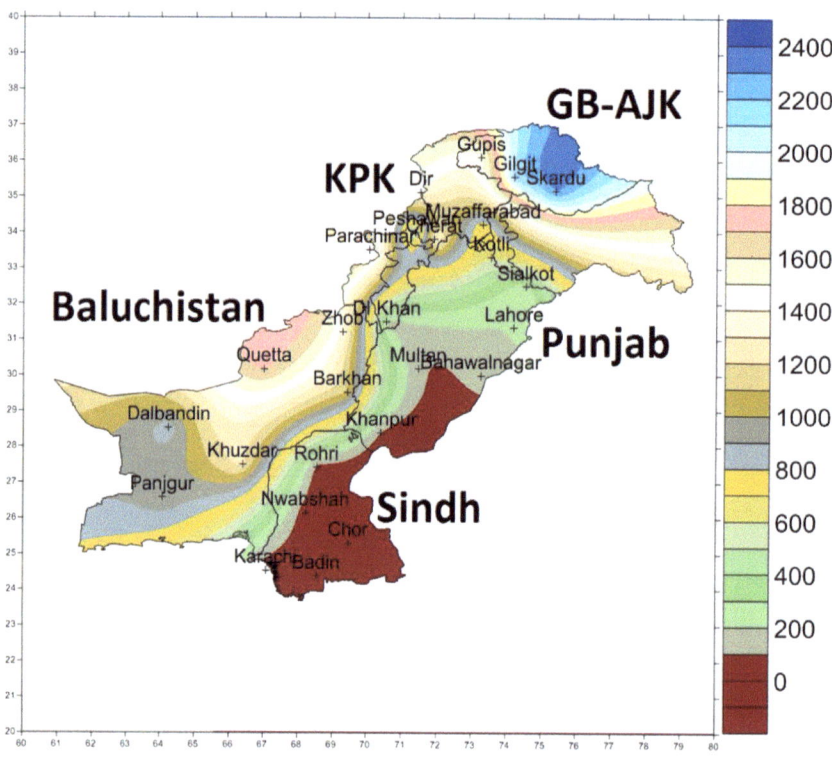

Figure 2.3: Province-wise selected stations with relief in meters.

3 Verification of GCMs and RCMs

GCMs are known to have coarse horizontal resolutions with feedback and simulation biases. RCMs, on the other hand, are more sophisticated and have high horizontal resolutions, yet are not free from biases since both regional downscaling simulation and the parent driving GCM induce spatial and temporal biases (see e.g., Burhan et al., 2014). For the purpose of projecting robust climate change extremes and to have high confidence in the results, model selection is made by gauging its ability to emulate observed climatology both for temperature and precipitation. Moreover, Taylor diagrams for GCMs and RCMs are made to obtain a concise statistical summary of how well the simulated climatology match the observed climatology in terms of their correlation, their root-mean-square difference and the ratio of their variances (Taylor, 2001).

3.1 GCMs emulating observed climatology

Area averaged mean annual maximum and minimum temperatures, and area averaged total annual rainfall is extracted from the 7 GCMs and put to comparison with the observed climatology of maximum and minimum temperature, and rainfall. Province based climatologies are constructed (Figure 3.1). It is seen that in all provinces, maximum and minimum temperature climatologies are well represented by all GCMs except for Had-GEM2-AO (which is not in fact the actual Hadley center simulated model, but its modeling center is NIMR, South Korea). In terms of precipitation, generally models tend to have high scatter, so the precipitation results for GCMs thereafter tend to remain in low confidence.

3.1.1 GCMs emulating observed climatology over Baluchistan

In terms of rainfall in Baluchistan, quite haphazard climatological patterns displayed by different GCMs suggests that the coarse resolution of the GCMs are unable to resolve rugged terrain and barren climate of the province. However, CCSM4 to some extent is able to capture the peaks of annual rainfall cycle, yet with considerable wet biases yielding over-estimated results for the past climate of the province.

In Baluchistan province, GCMs correlate well (95-99%), root-mean-square difference is small (less than 0.5 units), and the models lie within ±0.5 standard deviation of the observed maximum and minimum temperatures. Nevertheless, no good correlation is seen to be established between any of the models representing precipitation statistics of the Baluchistan province (Figure 3.2).

3.1.2 GCMs emulating observed climatology over GB-AJK

Both maximum and minimum temperatures in GB-AJK are well represented by simulated GCMs climatologies, however, with considerable cold biases. Moreover, GCMs tend to pick up the winter (DJFM) precipitation climatological cycle comparatively better than that in the summer (JJAS), yet, once again, with huge biases in the province. CCSM4 is seen as better representing the annual precipitation cycle, as compared to the rest of the models.

In GB-AJK, all the models except HadGEM2-AO are giving 95-99% correlation with the observed in terms of maximum temperature. All the models except HadGEM2-AO are within the half root mean square difference, as well as within the +0.5 standard deviation of the observed maximum temperature. Same results hold for the minimum temperature in GB-AJK. In terms of precipitation, there is both large variability and error, along with weak correlations (some models even giving negative correlations) in the GB-AJK Province. CCSM4 gives a comparatively better estimation with 60% correlation, however, both the root-mean-square difference and the ratio of the variances tends to remain on the higher side.

3.1.3 GCMs emulating observed climatology over KPK

Summer (JJAS) maximum and minimum temperature in KPK is better resolved as compared to that in the winter (DJFM) by GCMs. Precipitation climatologies represented by GCMs are poorly resolved for summer (JJAS) and partially better resolved in winters (DJFM), however, with both wet and dry biases. CCSM4, nevertheless, captures both winter (DJFM) and summer (JJAS) precipitation peaks, yet, once again suffers with dry biases.

Both simulated maximum and minimum temperatures show good correlations (95-99%), small root-mean-square differences, and within +0.5 standard deviation existence in the KPK province. There is however, the same exception for the HadGEM2-AO model which does not emulate the observed statistics at all. In terms of precipitation, the scatter of different GCMs is large with most of the models bearing poor Taylor Statistics. Nevertheless, CCSM4 model has a comparatively better correlation (nearly 80%), and within -0.5 standard deviation of the observed, yet the root-mean-square difference remains above 0.5 units in the KPK province.

3.1.4 GCMs emulating observed climatology over Punjab

Unlike, Baluchistan and GB-AJK, Punjab is a plane area with vast gradually altering slopes, hence resolving terrain borne climatic features of this province is comparatively less challenging for the GCMs. However, Punjab at the same time is the province that receives maximum amount of monsoon rainfall in JAS, and hence monsoon dynamics play an important role in determining the precipitation patterns and cycles of this region. Owing to this probable limitation, the GCMs have not been able to successfully emulate precipitation peaks of monsoon in JAS. Nevertheless, CCSM4 is the only model in our analysis that has shown comparatively better annual cycle representation of precipitation with the observed in the Punjab province.

The simulated maximum temperature in Punjab displays good correlations (90-99%) with the observed. The root-mean-square difference is small and the models tend to remain within an acceptable range of +0.5 standard deviation of the observed. This holds same for the minimum temperature. However, in terms of precipitation, only CCSM4 model displays high correlation (nearly 90%), a smaller root-mean-square difference and an acceptable standard deviation ratio with the observed climate of Punjab.

3.1.5 GCMs emulating observed climatology over Sindh

Maximum and minimum temperature climatology of Sindh is resolved within justifiable ranges. However, similar to other provinces, precipitation cycle remains a challenge to get resolved with currently analyzed GCMs' monsoon dynamics. CCSM4 captures the monsoon rainfall peaks better than the remaining models. However, quite large over-estimation of precipitation in CCSM4 in monsoon rains is also observed.

Simulated maximum temperature displays a high correlation (90-99%), and remains within an acceptable range of root-mean-square difference of the observed climate in Sindh province. In terms of minimum temperature, the correlation is even higher (models tending to be at 99% correlation). The ratio of the variances displayed by simulated minimum temperature is within ±0.5 units and the root-mean-square difference remains less than 0.5 units for the minimum temperature. In terms of precipitation bcc-CSM1.1 model displays a comparatively better correlation (nearly

99%), however, at the same time it suffers with high magnitudes of variance ratio, and high magnitudes of root-mean-square differences in Sindh province.

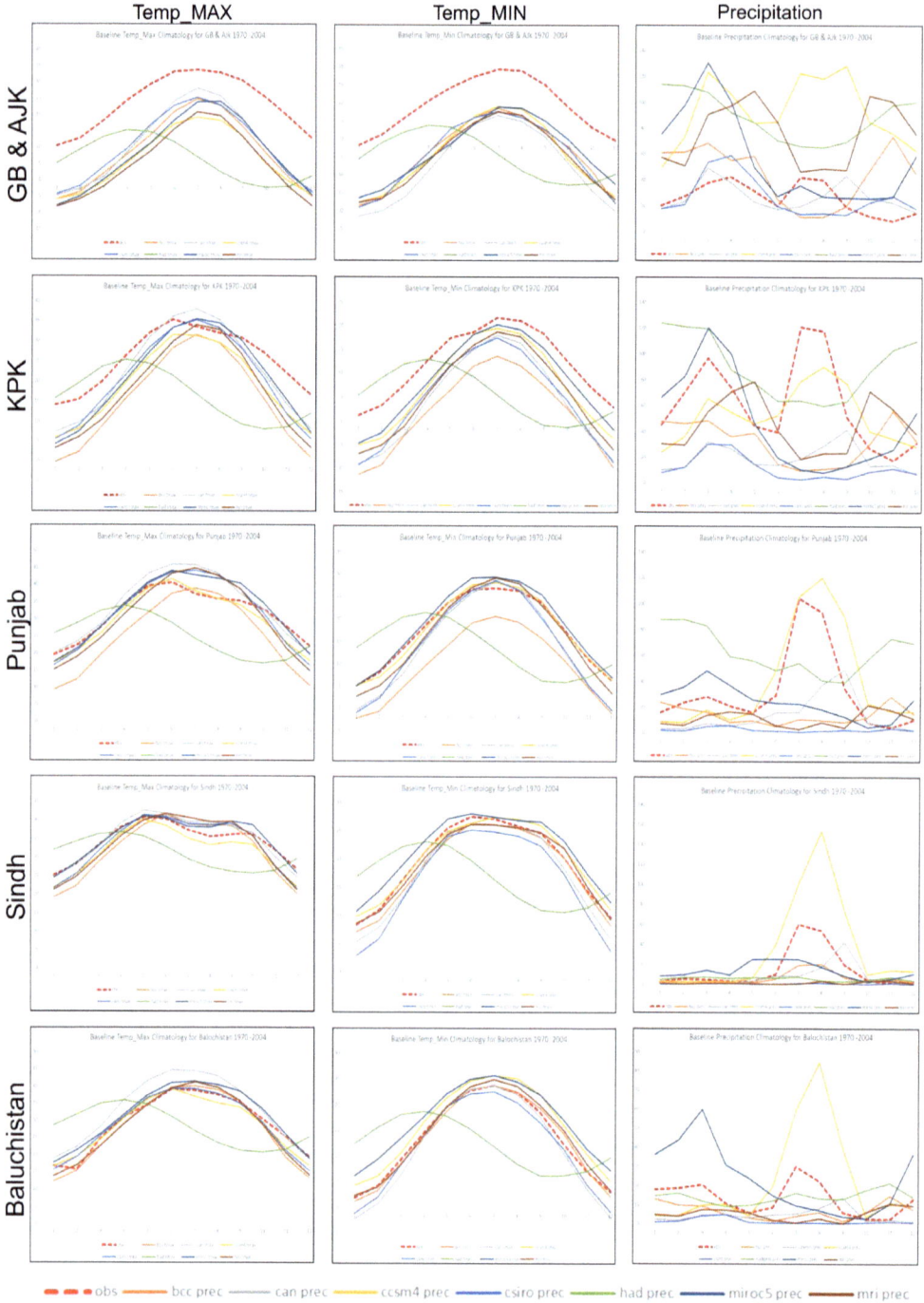

Figure 3.1: Observed and GCMs annual cycles of TMAX (°C), TMIN (°C) and Precip (mm) over 5 different provinces for 1970-2004 historical period.

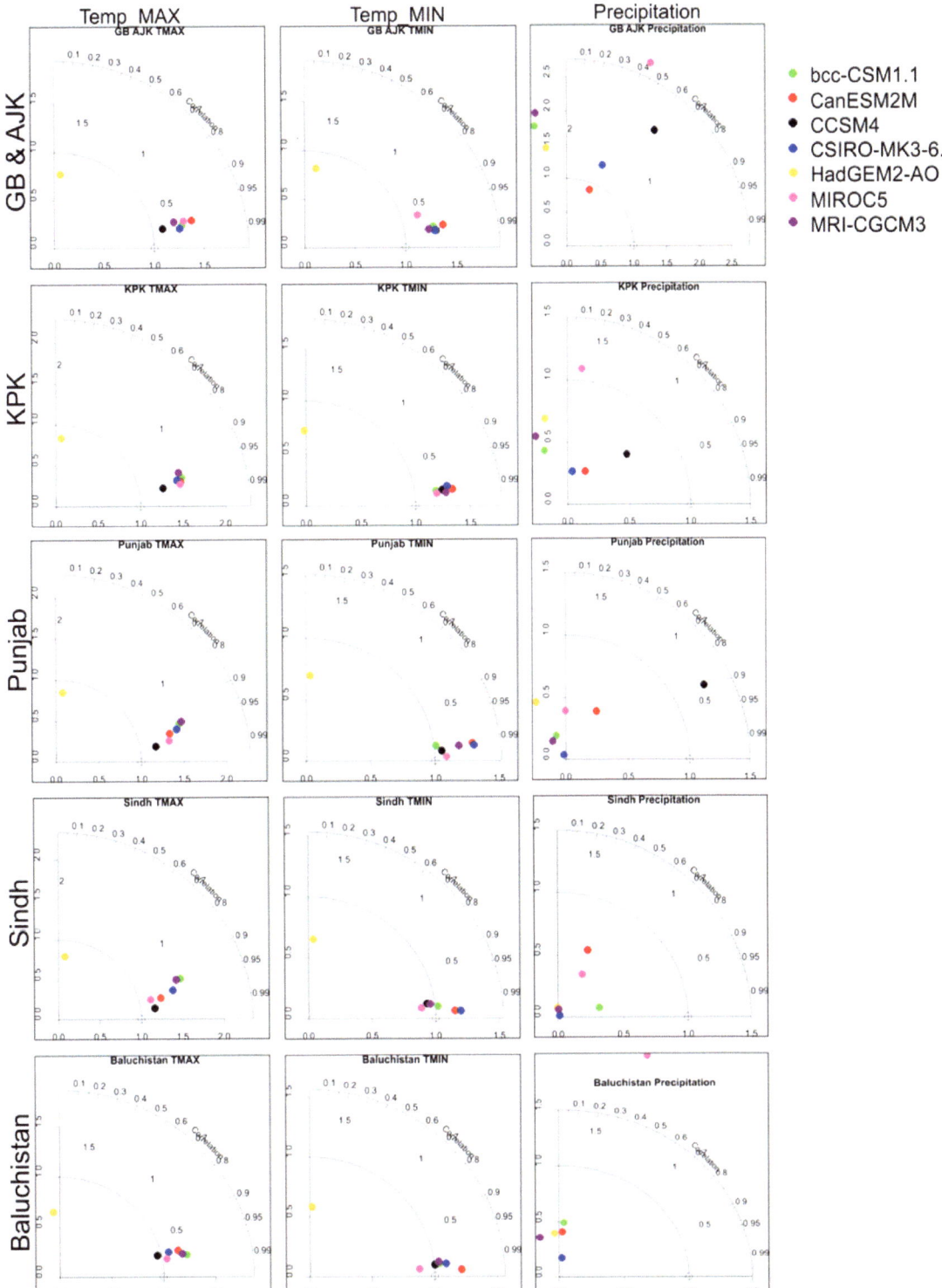

Figure 3.2: Province based Taylor Diagrams for TMAX, TMIN, and Precip, for the raw GCMs used in this study.

3.2 STATISTICALLY DOWNSCALED GCMs EMULATING OBSERVED CLIMATOLOGY

Statistically downscaled Nex-NASA GCMs output is area averaged for maximum temperature, minimum temperature, and precipitation independently over all the provinces of the country. Climatologies are extracted and annual cycles are constructed to make a comparison with the observed (Figure 3.3). Moreover, these climatologies are further put into Taylor Statistics to obtain high confidence over model selection (Figure 3.4). The results show good climatological aspects of Nex-NASA product both in terms of temperature and precipitation.

3.2.1 Statistically downscaled GCMs emulating observed climatology over KPK

Annual cycles of maximum and minimum temperatures, as well as of precipitation are well captured, however the temperatures suffer from cold biases whereas the precipitation suffers from dry biases. The winter peaks of precipitation are well overlapped with the observed, yet the summer peaks tend to remain below the observed, indicating under-estimation in the KPK province.

All parameters (TMAX, TMIN, and precipitation) display satisfying values for Taylor Statistics in KPK province. The correlations are high in precipitation (95-99%) whereas they are higher in TMAX and TMIN (> 99%) in the KPK region. The root mean square is less than 0.5 units for all the analyzed models and the parameters for the KPK region. Moreover, the root mean square difference is small (< 0.5 units) in precipitation, whilst it is further smaller in TMAX and TMIN.

3.2.2 Statistically downscaled GCMs emulating observed climatology over Punjab

The climatology of Punjab is brilliantly captured by the Nex-NASA product, for both maximum and minimum temperature, as well as for precipitation, for all the analyzed models. Monsoon peaks in precipitation cycle show small under-estimation by Can-ESM2 and CNRM-CM5, whereas they show small over-estimation by MRI-CGCM3, when compared with the observed. Nevertheless, the magnitude of under-estimation and over-estimation is quite small as compared to raw GCM products.

The temperature parameters display very small values for root-mean-square differences, as well as for the ratio of the variances, along with high correlations (> 99%) with the observed climate in Punjab. In terms of precipitation, the Nex-NASA product tends to display small ratios of variances (within ±0.5 standard deviation of observed), small root mean square differences (< 0.5 units), and simultaneous high correlations (> 95%) with the observed.

3.2.3 Statistically downscaled GCMs emulating observed climatology over Sindh

Both maximum and minimum temperatures display a nice overlapping of the Nex-NASA and the observed data over the Sindh province. Climatology of precipitation is also well captured, however, small under-estimation of precipitation in monsoon season is observed. Nevertheless, the biases are systematic and may be removed by simple bias correction techniques for the purpose of impact studies.

Temperature and precipitation climatologies of the Nex-NASA product are well coordinated with the observed, in terms of all three Taylor statistics over the Sindh region. The TMAX and TMIN have virtually no root mean square differences, and the ratios of the model variances to the observed is 1. Moreover the correlations are high (> 99%) in the region. In terms of precipitation, the root mean square differences are small (< 0.5 units). The ratios of variances to the observed tends to remain between 0.5 and 1. Furthermore, the correlations are high with values exceeding 99% for the region.

3.2.4 Statistically downscaled GCMs emulating observed climatology over Baluchistan

The Nex-NASA product in Baluchistan is well overlapped over observed climatology of the maximum and the minimum temperature. The precipitation cycle, though well coordinated, suffers from under-estimation in the summer season.

The TMAX and TMIN emulation of Nex-NASA product displays high correlations (99%) in the Baluchistan province. The root mean square difference is small (< 0.5 units), and the ratios of variances to the observed are between 0.5 and 1. In terms of precipitation, the correlations are high (80-95%), with ratios of variances to the observed ranging between 0.5 and 1 for the Baluchistan region.

3.2.5 Statistically downscaled GCMs emulating observed climatology over GB-AJK

Both maximum and minimum temperature as well as precipitation products of Nex-NASA display comparative annual cycles with the observed, yet the temperature is over-estimated whilst the precipitation is under-estimated in the GB-AJK region. Peaks for both winter and summer precipitation are captured nicely over the region,

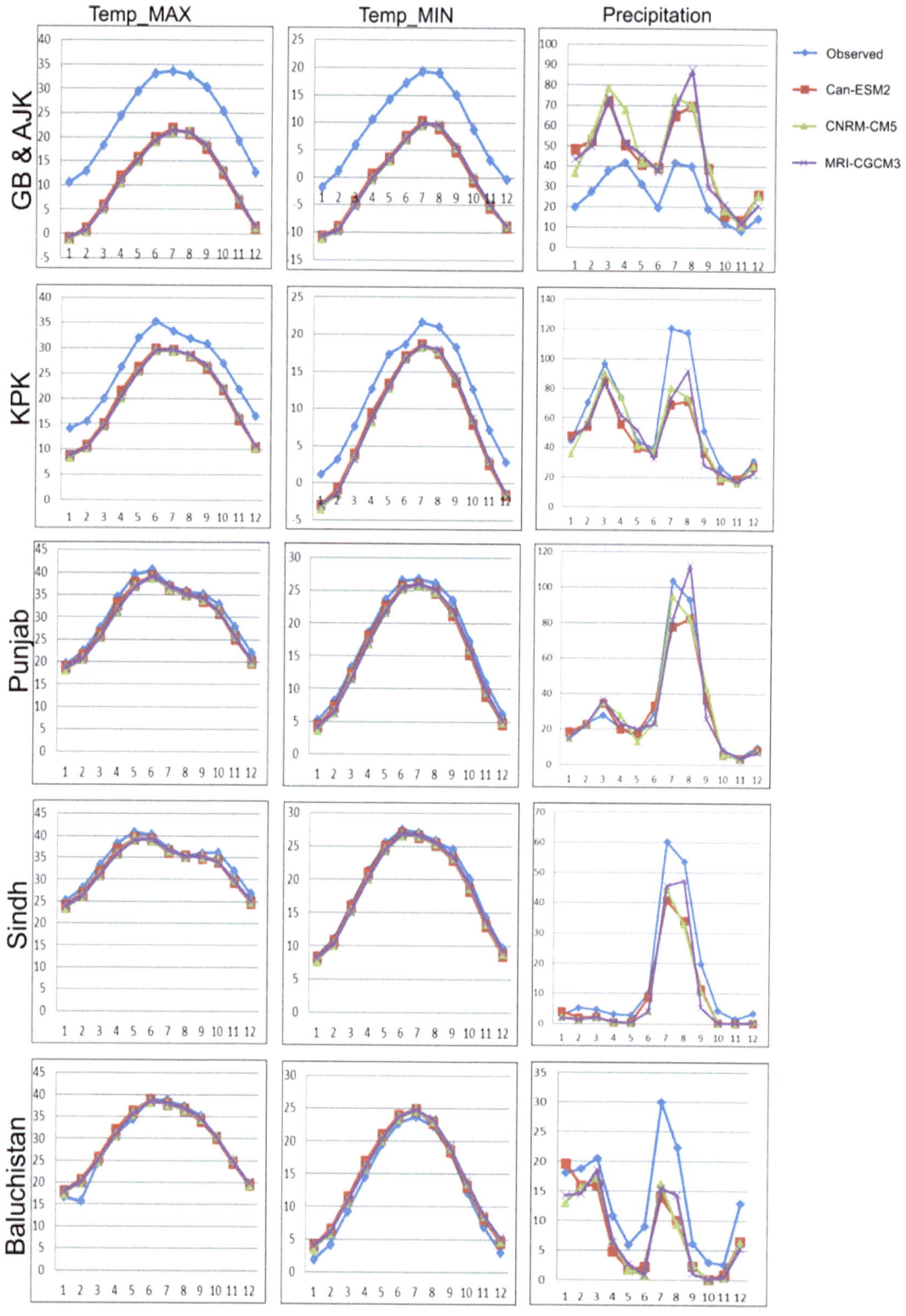

Figure 3.3: Observed and statistically downscaled Nex-NASA GCMs annual cycles of TMAX (°C), TMIN (°C) and Precip (mm) over 5 different provinces for 1970-2004 historical period.

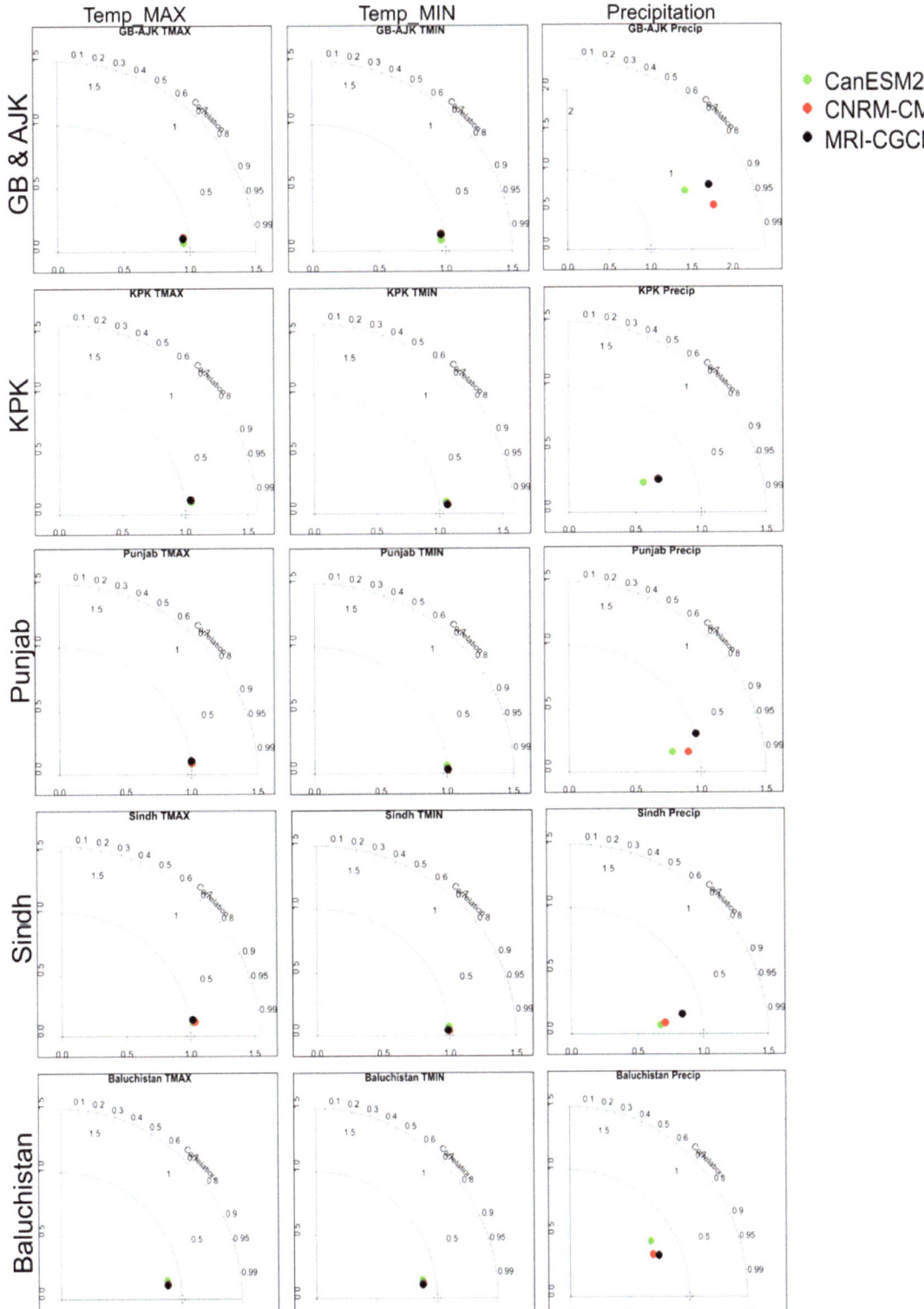

Figure 3.4: Province based Taylor Diagrams for TMAX, TMIN, and Precip, for the Nex-NASA GCMs used in this study.

however, the peak of the winter precipitation of the Nex-NASA data lags behind by one month leading to some possible time oriented biases in the region.

Taylor Statistics of maximum and minimum temperatures of GB-AJK display very good correlations (> 99%), whereas those of precipitation of GB-AJK display good correlations (90-95%) in the region. The ratios of variances in maximum and minimum temperatures are quite close to 1 (meaning almost identical variances), however, those in precipitation are between 1.5 and 2 (meaning large variations in variances) in the GB-AJK region. Moreover, root-mean-square differences in maximum and minimum temperatures are quite small, whereas those in precipitation are greater than 0.5 units.

3.3 Dynamically downscaled GCMs emulating observed climatology

Dynamical downscaling, in addition to its non-trivial initial and boundary conditions, features correction of terrain and relief borne climatic biases in the models. The CORDEX RCMs used in the study offers dynamically downscaled GCMs product over South Asian region. Annual climatologies from the six RCMs are extracted and plotted over observed to analyze credibility of the RCMs to emulate annual cycle over each province (Figure 3.5). It is important to note that RCMs are downscaling tools that may induce systematic or non-systematic errors in the output product, based on the driving GCM. The output product from an RCM therefore inherits two bias sources – one from the driving GCM, and the other from the RCM simulation.

3.3.1 Dynamically downscaled GCMs emulating observed climatology over GB-AJK

The CORDEX climatology for GB-AJK maximum and minimum temperature emulates the annual cycle in fact with huge systematic biases. Relatively better representation of the annual cycle is seen in SMHI RCA4 ICHEC-EC-EARTH RCM, yet suffers from small under-estimation in temperature output. The precipitation climatology of CORDEX RCMs in GB-AJK region displays large non-systematic errors that restrict the RCMs ability to correctly represent summer and winter precipitation peaks in the region. Nevertheless SMHI RCA4 ICHEC-EC-EARTH RCM is able to capture the winter precipitation with good ability, yet summer precipitation is quite over-estimated in the GB-AJK region.

Taylor Statistics of GB-AJK region displays all the CORDEX RCMs output for maximum and minimum temperatures to remain within 0.5-1.5 units of ratio of the variances, as well as within 0.5 units of root mean square differences to the observed. The correlations for TMAX and TMIN is > 99% and > 95% respectively. However, none of the CORDEX RCMs successfully imitate Taylor Statistics for observed precipitation in the GB-AJK region (Figure 3.6).

3.3.2 Dynamically downscaled GCMs emulating observed climatology over KPK

Maximum and minimum temperatures from CORDEX models display fairly good climatological patterns with the observed, yet with both over-estimated and under-estimated systematic biases present. In terms of precipitation, all CSIRO modeling center RCMs have fairly captured summer precipitation peaks (though with small under-estimation), while they have been un-successful in capturing the winter precipitation peak (rather showing a trough instead of a peak) for the KPK region. Winter precipitation peak is fairly well captured by REMO2009 MPI-ESM-LR RCM (with small under-estimation) over the KPK province.

The Taylor Statistics for KPK province displays good correlations (> 95 % for TMAX and > 99 % for TMIN) for all the CORDEX analyzed models. The root mean square differences of all the models are within 0.5 units of the observed, whilst the ratio of variances to the observed lie within 1-1.5 units for all the analyzed CORDEX models. For precipitation, REMO2009 MPI-ESM-LR and SMHI RCA4 ICHEC EC EARTH RCMs tend to have relatively high correlations of 50 % and 60 % respectively, with the observed. However, both models have high root mean square differences with the observed (> 0.5 units), in terms of KPK precipitation statistics.

3.3.3 Dynamically downscaled GCMs emulating observed climatology over Punjab

Both maximum and minimum temperature climatologies of CORDEX RCMs display good overlapping with the observed in Punjab province. In terms of precipitation all CSIRO modeling center RCMs have captured the winter precipitation peaks with good precision (yet they display early onset and gradual withdrawal of monsoon rains, as compared to the observed), however, they are not good at representing winter precipitation in Punjab province with correct figures. Nevertheless, REMO2009 MPI-ESM-LR RCM is able to capture the winter precipitation peak of Punjab province, however, it suffers from dry biases.

The root mean square differences of TMAX and TMIN of Punjab province from CORDEX RCMs are small (< 0.5 units) when compared with the observed. The ratio of the variances to the observed for TMAX and TMIN remains within 1-1.5 from all the analyzed CORDEX RCMs. The correlations are high (> 95 % for TMAX and > 99 % for TMIN) for all the CORDEX analyzed models in Punjab province. In terms of precipitation, correlations for all CSIRO based RCMs are high (> 80 %), yet the root mean square differences are > 0.5 units. Nevertheless, the ratio of the variances to the observed precipitation is quite close to 1 which means that the scatter of the CSIRO RCMs data is closed to that in observed over Punjab province.

3.3.4 Dynamically downscaled GCMs emulating observed climatology over Sindh

In Sindh province, both maximum and minimum temperature climatologies are well exhibited when compared with the observed. Precipitation however, is suffering from both systematic and time lead/lag biases from CORDEX RCMs. Relatively better representation of Sindh precipitation climatology is given by REMO2009 MPI-ESM-LR RCM where it captures monsoon peak with good precision but with small over-estimation in the Sindh province.

The Taylor Statistics of Sindh province for TMAX and TMIN from CORDEX RCMs display high correlations (> 95 % and > 99 % respectively), small root mean square differences (< 0.5 units), and the ratio of their variances to the observed are quite close to 1. The REMO2009 MPI-ESM-LR RCM seems to better imitate the observed precipitation statistics with > 90 % correlation, < 0.5 units of root mean square difference, and proximity of the ratio of the variances to one.

3.3.5 Dynamically downscaled GCMs emulating observed climatology over Baluchistan

The CORDEX RCMs are displayed to emulate maximum and minimum temperature climatologies of Baluchistan with good precision, yet there are small over-estimated and under-estimated biases. The precipitation climatologies for summer season are also well captured by the majority of the analyzed models (with systematic biases), however, winter season precipitation climatology is not captured by any of the CORDEX RCMs over Baluchistan.

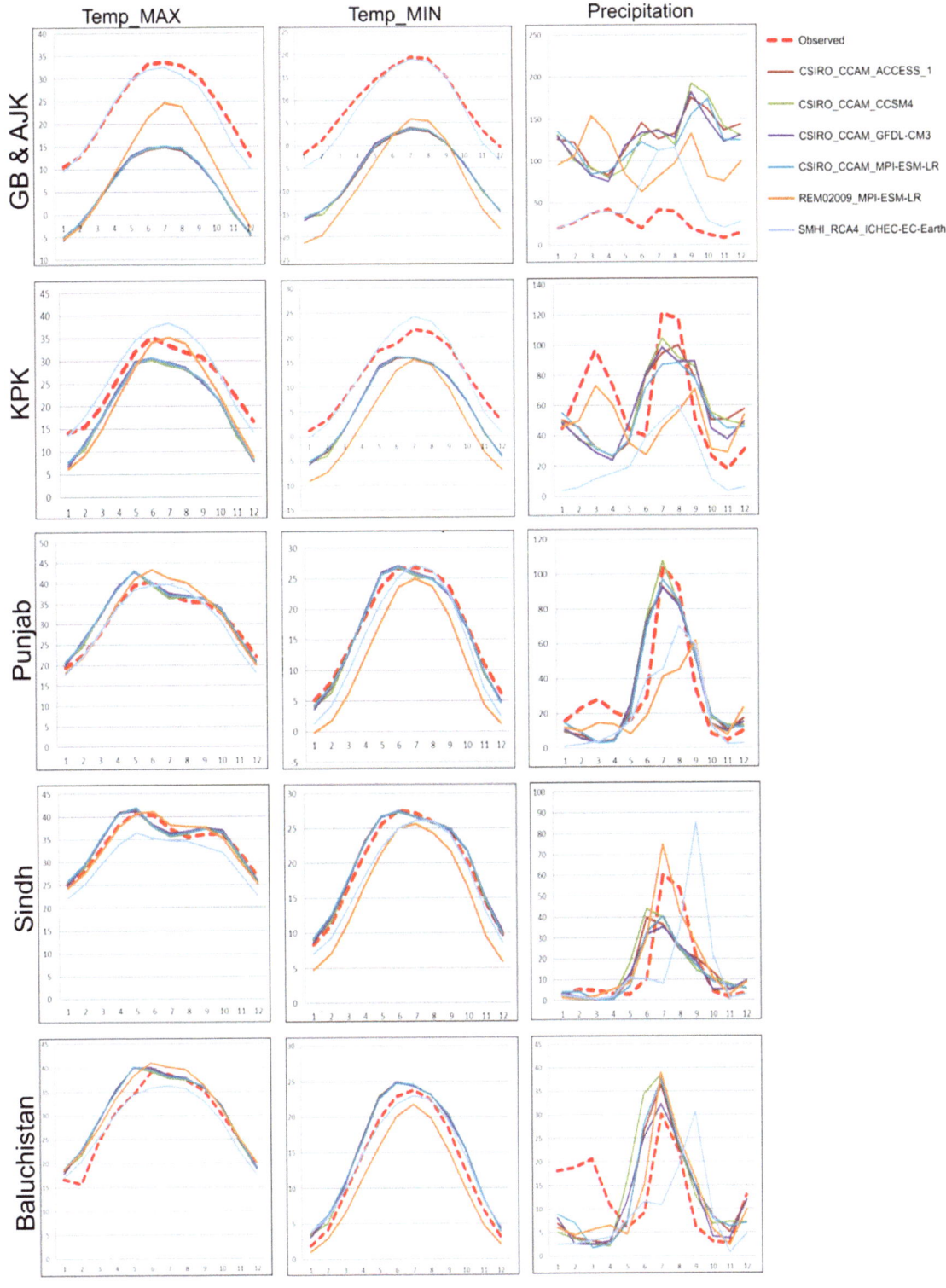

Figure 3.5: Observed and dynamically downscaled CORDEX GCMs annual cycles of TMAX (°C), TMIN (°C) and Precip (mm) over 5 different provinces for 1970-2004 historical period.

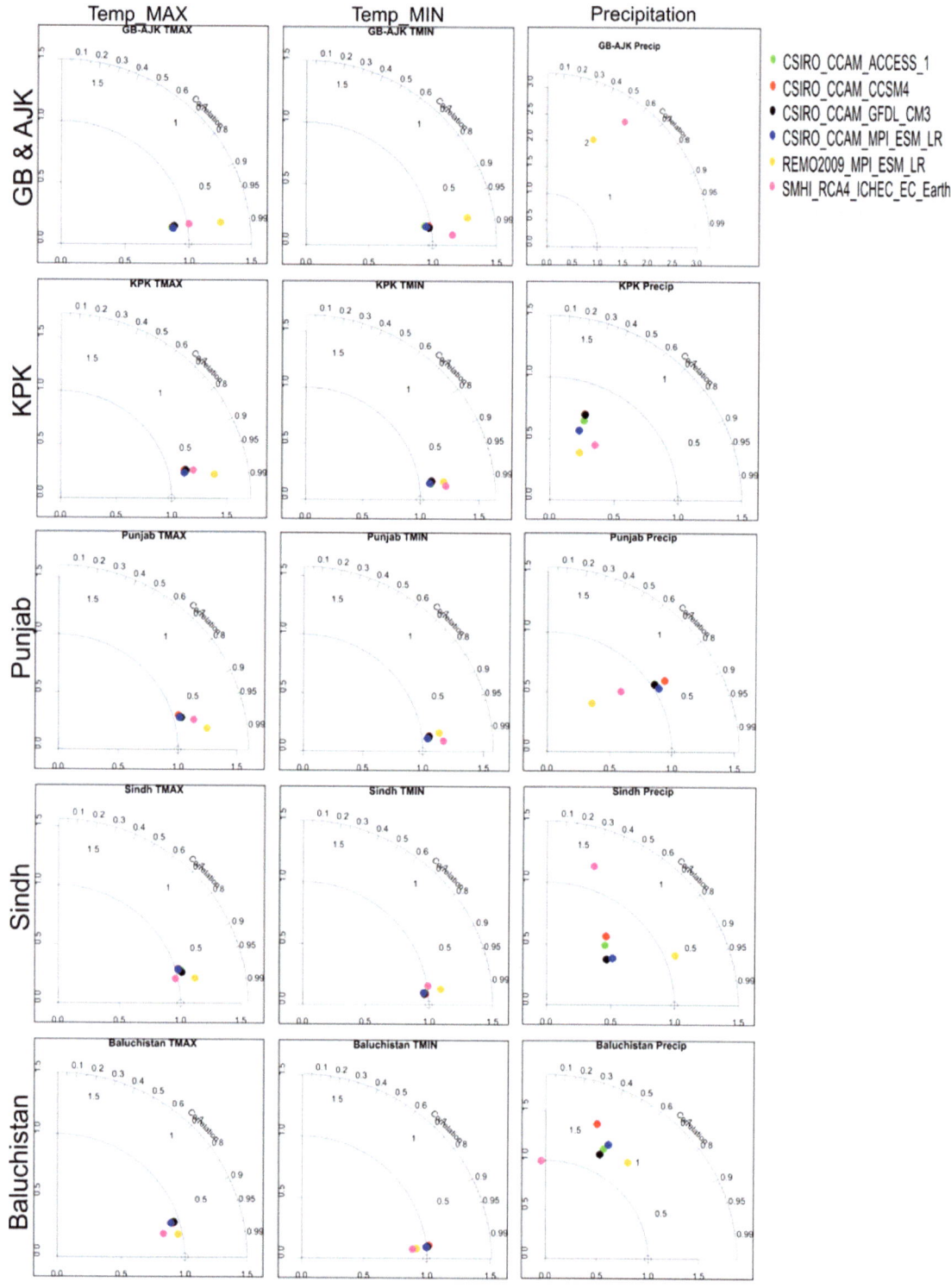

Figure 3.6: Province based Taylor Diagrams for TMAX, TMIN, and Precip, for the CORDEX RCMs used in this study.

The Taylor Statistics for TMAX and TMIN in Baluchistan province from all the analyzed CORDEX models display good correlations (> 95 % for TMAX and > 99 % for TMIN), as well as small root mean square differences (< 0.5 units). The ratio of the variances to the observed for TMAX and TMIN is close to 1 which indicates that CORDEX models tend to have nearly same spread in their climatological data as seen in that of observed data for TMAX and TMIN. As far as the precipitation is concerned, no robust Taylor Statistics could be found by any of the CORDEX RCMs for Baluchistan province.

3.4 Verdict over GCMs and RCMs subsets selection

Models' behavior in emulating climate of different provinces of Pakistan leads us to derive some generalized inferences:

a. Raw GCMs have generally shown only systematic biases in TMAX and TMIN, while precipitation is not resolved by any of the raw GCMs except by CCSM4. It is also worth mentioning that grid resolution alone, is not to be taken as a standard for a GCM to emulate observed climate. As is seen in Figure 3.7, CCSM4 has a coarser grid resolution than both the Australian based and the Japanese based models, yet it serves us well with resolving the regional climate better than any of the remaining analyzed models.

b. Since all Nex-NASA statistically downscaled GCMs are bias corrected, therefore they display highest confidence in emulating all the parameters correctly.

c. Individual representation of CORDEX RCMs in terms of describing the climate over different provinces induces huge systematic and non-systematic errors, to which the solution lies in taking the ensemble average of all the RCMs for further analysis. Moreover, it is seen that all RCMs from CSIRO modeling center have shown quite similar climatologies and statistics which leads to a hypothesis that the CSIRO models have large influence of the RCM (CCAM) from which they are downscaled, rather than that of the driving GCMs.

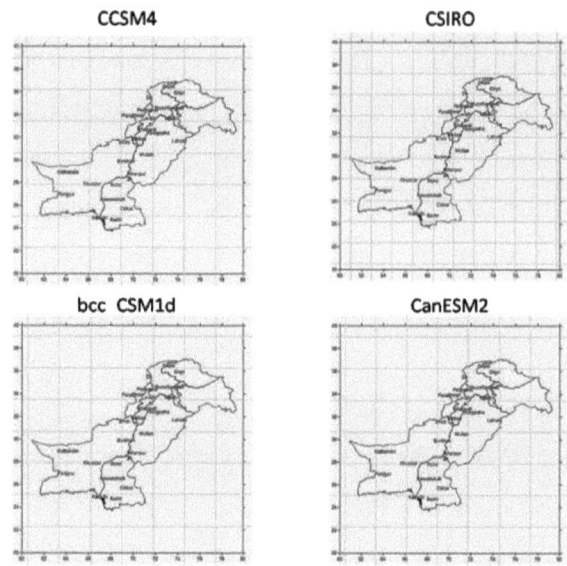

GCM\Province	GB-AJK	KPK	Punjab	Sindh	Baluchistan
CCSM4	4	6	9	6	12
CSIRO-MK3.6.0	16	12	16	12	25
MIROC5	20	25	25	16	48
bcc-CSM1.1	4	6	9	6	12
Can-ESM2	4	6	9	6	12

Figure 3.7: GCM grids and province-wise grids count for selected GCMs.

Based on the ability of models to emulate observed climatologies and statistics (as described in the above discussion), the selected raw GCMs, the statistically downscaled GCMs, and the selected dynamically downscaled GCMs, are summarized in Table 3.1.

Table 3.1: Summary of the selected raw GCMs, the statistically downscaled GCMs, and the dynamically downscaled GCMs.

	GB/AJK	KPK	Punjab	Sindh	Baluchistan
Ensembles of RAW GCMs					
Precipitation	CCSM4	CCSM4	CCSM4	CCSM4	CCSM4
Temp Max	CCSM4 CSIRO-MK3.6.0 bcc-CSM1-1	CCSM4 CSIRO-MK3.6.0 MIROC5	CCSM4 MIROC5 CanESM2	CCSM4 MIROC5 CanESM2	CCSM4 MIROC5 CSIRO-MK3.6.0
Temp Min	MIROC5 bcc_CSM1 CSIRO-MK3.6.0	CCSM4 MIROC5 bcc-CSM1-1	CCSM4 bcc-CSM1-1 MIROC5	CCSM4 MIROC5 bcc-CSM1-1	CCSM4 CSIRO-MK3.6.0 bcc-CSM1-1
Ensembles of Statistically Downscaled Nex-NASA GCMs					
Precipitation	CAN-ESM2 CNRM-CM5 MRI-CGCM3	CAN-ESM2 CNRM-CM5 MRI-CGCM3	CAN-ESM2 CNRM-CM5 MRI-CGCM3	CAN-ESM2 CNRM-CM5 MRI-CGCM3	CAN-ESM2 CNRM-CM5 MRI-CGCM3
Temp Max	CAN-ESM2 CNRM-CM5 MRI-CGCM3	CAN-ESM2 CNRM-CM5 MRI-CGCM3	CAN-ESM2 CNRM-CM5 MRI-CGCM3	CAN-ESM2 CNRM-CM5 MRI-CGCM3	CAN-ESM2 CNRM-CM5 MRI-CGCM3
Temp Min	CAN-ESM2 CNRM-CM5 MRI-CGCM3	CAN-ESM2 CNRM-CM5 MRI-CGCM3	CAN-ESM2 CNRM-CM5 MRI-CGCM3	CAN-ESM2 CNRM-CM5 MRI-CGCM3	CAN-ESM2 CNRM-CM5 MRI-CGCM3
Ensembles of Dynamically Downscaled CORDEX RCMs					
Precipitation	SMHI-RCA4-ICHEC-EC-EARTH	CSIRO-CCAM-ACCESS-1 CSIRO-CCAM-CCSM4 CSIRO-CCAM-GFDL-CM3 CSIRO-CCAM-MPI-ESM-LR REMO2009-MPI-ESM-LR SMHI-RCA4- ICHEC-EC-EARTH	CSIRO-CCAM-ACCESS-1 CSIRO-CCAM-CCSM4 CSIRO-CCAM-GFDL-CM3 CSIRO-CCAM-MPI-ESM-LR REMO2009-MPI-ESM-LR SMHI-RCA4- ICHEC-EC-EARTH	REMO2009-MPI-ESM-LR	CSIRO-CCAM-ACCESS-1 CSIRO-CCAM-CCSM4 CSIRO-CCAM-GFDL-CM3 CSIRO-CCAM-MPI-ESM-LR REMO2009-MPI-ESM-LR
Temp Max	SMHI-RCA4-ICHEC-EC-EARTH	CSIRO-CCAM-ACCESS-1 CSIRO-CCAM-CCSM4 CSIRO-CCAM-GFDL-CM3 CSIRO-CCAM-MPI-ESM-LR REMO2009-MPI-ESM-LR SMHI-RCA4- ICHEC-EC-EARTH	CSIRO-CCAM-ACCESS-1 CSIRO-CCAM-CCSM4 CSIRO-CCAM-GFDL-CM3 CSIRO-CCAM-MPI-ESM-LR REMO2009-MPI-ESM-LR SMHI-RCA4- ICHEC-EC-EARTH	CSIRO-CCAM-ACCESS-1 CSIRO-CCAM-CCSM4 CSIRO-CCAM-GFDL-CM3 CSIRO-CCAM-MPI-ESM-LR REMO2009-MPI-ESM-LR SMHI-RCA4-ICHEC-EC-EARTH	CSIRO-CCAM-ACCESS-1 CSIRO-CCAM-CCSM4 CSIRO-CCAM-GFDL-CM3 CSIRO-CCAM-MPI-ESM-LR REMO2009-MPI-ESM-LR SMHI-RCA4-ICHEC-EC-EARTH
Temp Min	SMHI-RCA4-ICHEC-EC-EARTH	CSIRO-CCAM-ACCESS-1 CSIRO-CCAM-CCSM4 CSIRO-CCAM-GFDL-CM3 CSIRO-CCAM-MPI-ESM-LR REMO2009-MPI-ESM-LR SMHI-RCA4- ICHEC-EC-EARTH	CSIRO-CCAM-ACCESS-1 CSIRO-CCAM-CCSM4 CSIRO-CCAM-GFDL-CM3 CSIRO-CCAM-MPI-ESM-LR REMO2009-MPI-ESM-LR SMHI-RCA4- ICHEC-EC-EARTH	CSIRO-CCAM-ACCESS-1 CSIRO-CCAM-CCSM4 CSIRO-CCAM-GFDL-CM3 CSIRO-CCAM-MPI-ESM-LR REMO2009-MPI-ESM-LR SMHI-RCA4-ICHEC-EC-EARTH	CSIRO-CCAM-ACCESS-1 CSIRO-CCAM-CCSM4 CSIRO-CCAM-GFDL-CM3 CSIRO-CCAM-MPI-ESM-LR REMO2009-MPI-ESM-LR SMHI-RCA4-ICHEC-EC-EARTH

4 Types of Climate Indices

After scrutinizing the models and their selection, province based core climate indices are calculated (Table 4.1). We have calculated and analyzed 25 indices – 15 indices related to the temperature and 10 indices related to the precipitation. Indices are driven from maximum temperature, minimum temperature, and precipitation. The climate indices are of five different types:

4.1 Percentile-Based Indices

Occurrence of cold nights, occurrence of warm nights, occurrence of cold days, occurrence of warm days, very wet days, and extremely wet days, are all percentile based indices.

4.2 Absolute Indices

These indices represent maximum or minimum values within a season or a year. Maximum of daily maximum temperature, maximum of daily minimum temperature, minimum of daily maximum temperature, minimum of daily minimum temperature, maximum 1-day precipitation amount, and maximum 5-day precipitation amount, are all absolute indices.

4.3 Threshold Indices

The indices are those in which the number of days on which a temperature or precipitation value falls above or below a fixed threshold. Annual occurrence of frost days, annual occurrence of ice days, annual occurrence of summer days, annual occurrence of tropical nights, number of heavy precipitation days, and number of very heavy precipitation days, are all threshold indices.

4.4 Duration Indices

These indices signify periods of excessive warm, cold, wetness or dryness in the climate. Cold spell duration indicator, warm spell duration indicator, consecutive dry days, and consecutive wet days, are all duration indices.

4.5 Societal Impact indices

The indices could have significant societal impacts. Annual precipitation total, diurnal temperature range, simple daily intensity index, and annual contribution from very wet days, all fall into societal impact indices.

Table 4.1: Core climate indices calculated and analyzed in this study.

S. NO.	INDEX	INDEX LONG NAME	INDEX DETAIL	UNITS
1	FD	Number of frost days	Annual count of days when TN (daily minimum temperature) < 0oC	Days
2	SU	Number of summer days	Annual count of days when TX (daily maximum temperature) > 25oC	Days
3	ID	Number of icing days	Annual count of days when TX (daily maximum temperature) < 0oC	Days
4	TR20	Number of tropical nights	Annual count of days when TN (daily minimum temperature) > 20oC	Days
5	TXx	Monthly maximum value of daily maximum temperature	The maximum of daily maximum temperature for each month	°C
6	TNx	Monthly maximum value of daily minimum temperature	The maximum of daily minimum temperature for each month	°C
7	TXn	Monthly minimum value of daily maximum temperature	The minimum of daily maximum temperature for each month	°C
8	TNn	Monthly minimum value of daily minimum temperature	The minimum of daily minimum temperature for each month	°C
9	TN10p	Cool nights	Percentage of days when TN < 10th percentile	Days
10	TX10p	Cool days	Percentage of days when TX < 10th percentile	Days
11	TN90p	Warm nights	Percentage of days when TN > 90th percentile	Days
12	TX90p	Warm days	Percentage of days when TX > 90th percentile	Days
13	WSDI	Warm spell duration indicator	Annual count of days with at least 6 consecutive days when TX > 90th percentile	Days
14	CSDI	Cold spell duration indicator	Annual count of days with at least 6 consecutive days when TN < 10th percentile	Days
15	DTR	Diurnal temperature range	Monthly mean difference between TX and TN	°C
16	RX1day	Max 1-day precipitation amount	Monthly maximum 1-day precipitation	mm
17	Rx5day	Max 5-day precipitation amount	Monthly maximum consecutive 5-day precipitation	mm
18	SDII	Simple pricipitation intensity index	Annual total precipitation divided by the number of wet days (defined as PRCP>=1.0mm) in the year	mm/day
19	R10	Number of heavy precipitation days	Annual count of days when PRCP>=10mm	Days
20	R20	Number of very heavy precipitation days	Annual count of days when PRCP>=20mm	Days
21	CDD	Consecutive dry days	Maximum number of consecutive days with RR<1mm	Days
22	CWD	Consecutive wet days	Maximum number of consecutive days with RR>=1mm	Days
23	R95p	Very wet days	Annual total PRCP when RR>95th percentile	mm
24	R99p	Extremely wet days	Annual total PRCP when RR>99th percentile	mm
25	PRCPTOT	Annual total wet-day precipitation	Annual total PRCP in wet days (RR>=1mm)	mm

RClimDex is employed for index calculation in this study. The RClimDex provides a friendly graphical user interface to compute 27 core indices (see e.g., Karl et al., (1999); Peterson et al., (2001); Zhang et al. (2005)). It also conducts simple quality control on the input daily data. It has been developed and maintained by Xuebin Zhang and Yang Feng at Climate Research Division.

5 RESULTS AND DISCUSSION OF OBSERVED CLIMATE INDICES

Province based analysis of observed climate indices suggests some substantial increases and decreases in the trends over the baseline period. Station based analysis indicates significant trend changes at some stations while insignificant trend changes at others over each province. Although climatic features are not bounded by administerial or political boundaries, yet the selected administerial bounded regions somewhat describes distinct climatic features and hence are used for province based analysis.

5.1 Observed Indices of GB-AJK

Upon the analysis of Cold spell duration indicator it is seen that no significant change has occurred over the past 54 years (Figure 5.1). General trend is increasing but is not statistically significant. Consecutive wet days in AJK and GB has displayed a contrast in the trend observation of the index. The stations at GB (Gilgit, Gupis and Skardu) tend to display an insignificant increasing trend whilst the stations at AJK tend to show significantly decreasing trend. The decreasing trend of consecutive wet days in Kotli is six days while that in Muzaffarabad is three days with confidence at 99% level. In AJK and GB, the DTR show statistically significant increasing trend over the past 54 years. In GB, Gilgit displays a DTR trend increase of 2.2°C, Gupis displays that of 3°C, and Skardu displays that of 3.1°C over the past 54 years. In AJK, Kotli displays a DTR trend increase of 1.2°C, and Muzaffarabad displays that of 1°C over the past 54 years. A general increasing trend is observed in the analysis of frost days trend in AJK and GB. In GB, Gupis displays a frost days increasing trend of 21 days with 98% confidence, and Skardu displays a frost days increasing trend of 10 days with confidence of 92% over the 1960-2013 past 54 years period. Gilgit displays no change in its trend of frost days index. In AJK, the trend tends to be increasing but with smaller confidence and smaller number of days increasing over the historical period.

Ice days index when TMAX is less than a temperature threshold, indicates a general decreasing trend in AJK and GB. In GB, the ice days in Gilgit has decreased by 19 days, in Gupis by 21 days, and in Skardu by 30 days at 99% level of significance over the historical period of 1960-2013. In AJK, no significant trend is observed over the analysis of ice days index in past data. Heavy precipitation days for GB have been identified as those in which PR > 10 mm, while heavy precipitation days for AJK have been identified as those in which PR > 20 mm. In case of GB, Gupis displays a statistically significant (at 99% level) increasing trend of 6 days, while Skardu displays an increasing trend of 4 days at 93% level over the past 54 years. Giglit displays no significant trend in the number of heavy precipitation days over the historical analysis. In case of AJK, Kotli displays a statistically significant increasing trend of the heavy precipitation days with an increment of 6 days, and Muzaffarabad displays a statistically significant increasing trend of the heavy precipitation days with an increment of 18 days over the historical period of 1960-2013. Very wet days precipitation when annual total PRCP>95th percentile displays generally an increasing trend in AJK and GB over the past 54 years. Statistically significant increasing trend at 99% level is observed in Kotli and Muzaffarabad which shows that the annual total precipitation for very wet days has increased by 225 mm and by 432 mm respectively over the historical 1960-2013 historical period. No significant trend is observed in regions of GB probably owing to some very heavy precipitation years taken inclusive.

Monthly maximum consecutive 5-day precipitation amount in stations of AJK have shown significant increase with 48 mm in Kotli (at 90% level) and with 96 mm in Muzaffarabad (at 99% level). However, over GB, no significant change in the monthly maximum consecutive 5-day precipitation amount is identified upon the analysis of the historical data. Increase in 5-day consecutive precipitation may lead to Glacier lake outburst flooding in the Northern areas of Pakistan. Summer days when annual count of daily maximum temperature >25°C displays an increasing trend in AJK and GB. In GB, Gilgit displays an increasing trend of 24 days with 99% level of confidence, Gupis displays an increasing trend of 15 days with 95% level of confidence, and Skardu displays an increasing trend of 27 days with 99% level of confidence over the past 54 years. In AJK, Kotli displays an increasing trend of 13 days with 90% level of confidence, and Muzaffarabad displays an increasing trend of 17 days with 95% level of confidence. In AJK and GB, there is an increasing trend in the number of cool nights in all the stations taken, however none of the stations give significance at 95% level. In AJK and GB, a decreasing trend of warm nights is observed over the analysis of the 1960-2013 historical period. The decreasing trend is significant at 99% level for Gilgit (7 days decrease), Gupis (17 days decrease), and Kotli (9 days decrease). No significant trend is observed in Muzaffarabad and Skardu in the analysis of warm nights.

Tropical nights calculated by setting annual count when TN(daily minimum)>20°C display decreasing trends in the AJK and GB. The decreasing trend of GB is significant at 99% level. In GB, Gilgit displays a rate of decrease of 16 nights per 53 years, Gupis displays a rate of decrease of 6 nights per 53 years, and Skardu displays a rate of decrease of 8 nights per 53 years. There is no significant trend displayed by Kotli and Muzaffarabad. In GB, there is a significant decrease in the trend of cool days. In Gilgit, there is a decrease of 7 days at 99% level, in Gupis, there is a decrease of 5 days at 90% level, and in Skardu, there is a decrease of 12 days at 99% level. No significant change in the trends of cool days in Kotli and Muzaffarabad is observed in the past 54 years. Warm days calculated by the percentage of days when TX>90th percentile displays an increasing trend in AJK and GB over the analysis of the 1960-2013 historical period. In Gilgit, there is a significant increasing trend of 11 days at 99% level, in Gupis, there is a significant trend of 7 days at 95% level, and in Skardu, there is a significant increasing trend of 13 days at 99% level over the past 54 years. In AJK, Muzaffarabad displays a significant increasing trend of 9 warm days with confidence of 99% over the past data analysis. Kotli, however, displays no significant trend observation in case of warm days index. There is a general increasing trend in the warm spell duration indicator over AJK and GB. Skardu in GB show relatively higher confidence in displaying 11 days increase in the warm spell duration index over the past 54 years. The index is of vital significance since major GLOF events are effects of prolonged heat waves over the region.

5.2 Observed indices of KPK

In KPK, cold spell duration index has a general decreasing trend which means that cold waves have decreased over the past in KPK. In DI Khan and Peshawar, a linear trend decrease of 15 days (significant at 95% level) and 7 days (significant at 93% level) respectively in the cold spell duration index, is observed upon the analysis of past 54 years data (Figure 5.2). Slope estimate of Dir is not statistically significant, while Parachinar displays a statistically significant increasing trend of 33 days in the cold spell duration indicator upon the analysis of past 54 years data. No statistically significant trends are observed in consecutive wet days in KPK over the 1960-2013 past data analysis. Diurnal temperature range in KPK show clear signals increasing trends over the past 54 years data analysis. The stations at Dir, Parachinar, and Cherat show high confidence in increasing trends with 2.2°C, 3.1°C, and 2.5°C respectively at 95 %

level. On the other hand, diurnal temperature ranges of DI Khan and Peshawar show a decreasing trend but with smaller confidence and smaller statistical significance. Upon the analysis of frost days in KPK, the annual count of days with minimum temperature < 0.0°C taken as a threshold, gives no significant trend over the analysis of the past 54 years data. However, one station, Parachinar gives a statistically significant (at 99 % level) trend of 62 days increase in the frost days index. This substantial increase is displayed in Parachinar after 1980 which continues to remain so till 2013. Also the frost days with specified thresholds, show significant trends in KPK for the past 54 year data analysis. Frost days in Parachinar with minimum temperature below -4°C show a significant increasing trend at 99 % level with slope estimate of 65 days increase over the past 54 years data. On the other hand, DI Khan and Peshawar show statistically significant decreasing trends of 16 days and 14 days respectively, both at 99% level of confidence. Dir and Cherat show no statistically significant change in the trends of frost days index.

In the past data analysis, annual total precipitation amount of DI Khan and Peshawar has significantly increased in KPK. In DI Khan, the total precipitation has increased by 148 mm, and in Peshawar, the total precipitation has increased by 263 mm, over the past 54 years (significance of trend is at 99% level). However, there is no significant trend observed in total precipitation of Dir, Parachinar and Cherat over the last 54 years. Heavy precipitation extremes of rainfall > 10 mm display significant increase in the number of days in DI Khan and Peshawar, with an increase of 4 days (significant at 95 % level) and an increase of 9 days (significant at 99 % level) over the past 54 years. Dir, Parachinar and Cherat show high variability with no significant trend observed in the number of days with heavy precipitation events. Very heavy precipitation events defined by annual count of days when PRCP>=20mm have significantly increased in DI Khan and Peshawar with slope estimates of 3 days and 6 days respectively, both significant at 99 % level of confidence. Once again Dir, Parachinar and Cherat does not show any significant trend in the very heavy precipitation days over the past 54 years data analysis. Annual total precipitation of very wet days when rainfall > 95th percentile displays a significant increasing trend in DI Khan and Peshawar of KPK. There is 91 mm increase in the total precipitation of very wet days in DI Khan (95% confidence level), and 123 mm increase in the total precipitation of very wet days in Peshawar (99% confidence level) over the analysis of past 54 years data. No significant trend in the annual total precipitation of very wet days is observed in the past data analysis of Dir, Parachinar and Cherat. Monthly maximum 5 day consecutive rainfall trend displays significant trend with 82 mm increase in Dir (at 95 % level) and with 61 mm increase in Peshawar (at 95 % level) over the analysis of past data over KPK. No statistically significant trend is observed in Parachinar, DI Khan and Cherat in the monthly maximum 5 day consecutive rainfall trend over the past data analysis.

Summer days index when the maximum temperature > 25°C displays a statistically significant trend towards increasing in Peshawar city of KPK province over the past 54 years data analysis. The trend shows an increase of 23 days in the past 54 years in Peshawar with significance at 99 % level. However, no other station of KPK show statistically significant trend in the summer days index. Summer days index with maximum temperature > 32°C has high confidence in the trend of Dir in KPK. Summer days index trend in Dir has displayed an increase of 41 days over the past 54 years data analysis. The confidence over the increasing trend in Dir is at 99 % level. Mean of maximum temperature over the past 54 years data display a high confidence in the trends of Dir and Peshawar of KPK. Dir displays 1.5°C increase, while Peshawar displays 0.9°C increase in the mean of the maximum temperature over the last 54 years. Trend of the mean of the minimum temperature over Parachinar in KPK show a significant decreasing trend of 3.3°C, while that over Peshawar in KPK show a significant increasing trend 0.9 of°C over the past 54 years data analysis. Both the trends are significant at 99 % level of confidence. Cool nights in Parachinar of KPK display an increasing trend of 20 nights with significance at 99 % level. On the other hand both DI Khan and Peshawar display decreasing trends of 5 nights and 6 nights significant over 95 % and 99 % level in the past 54 years. No significant trend is displayed by Dir in the trend of cool nights.

General trend of warm nights is decreasing in KPK with the exception of Peshawar whose trend is significantly increasing over the past 54 years. The rate of increase in 54 years is 8 nights with confidence at 99 % level. DI Khan on the other hand displays a significant decreasing trend of 4 nights in the past 54 years over KPK. Tropical nights in Dir and Cherat display significant decreasing trend while those in DI Khan and Peshawar display significant increasing trends. Tropical nights trend in Dir displays 19 nights decrease at 95 % level, and that in Cherat displays 42 nights decrease at 95 % level over the past 54 years. On the other hand, tropical nights trend in DI Khan displays

14 nights increase at 95 % level, and that in Peshawar displays 18 nights decrease at 99 % level over the past 54 years. There is a general decreasing trend observed in the cool days index of KPK over the past 54 years. Statistically significant trend is observed in Peshawar with 3 days increase in the cool days index with 99 % level of confidence. The general trend of warm days in KPK is increasing over the past 54 years data analysis. The rate of change is significant in Dir of KPK which displays an increase of 14 days over the past 54 years with confidence level at 99 % over the past 54 years. Warm spell duration index in KPK show generally an increasing trend which identifies the occurrences of heat waves in the region. Statistically significant increasing trends of 15 days in Dir and of 11 days in Parachinar are observed in the past 54 years data analysis of warm spell duration indicator in KPK.

5.3 OBSERVED INDICES OF PUNJAB

Cold spell duration indicator displays a general decreasing trend in Punjab over the analysis of past 54 years (Figure 5.3). Cold spells have decreased in Islamabad by 17 days at 99 % level, by 15 days in Sialkot at 90 % level, by 42 days in Lahore at 99 % level, by 19 days in Multan at 99 % level, by 33 days in Bahawalnagar at 99 % level, and by 13 days in Khanpur at 90 % level. Consecutive wet days in Punjab displays no significant trend, however Khanpur displays a statistically significant increase of 1 day (at 95 % level) in the maximum number of consecutive days with precipitation greater than 1 mm over the past 54 years analysis. DTR in Punjab displays a clear signal of decreasing trend in all the stations with statistically significant decreases of 1.5°C in Sialkot (at 99 % level), of 3.1°C in Lahore (at 99 % level), of 2.1°C in Multan (at 99 % level), and of 1.7°C in Khanpur (at 99 % level), over the past 54 years. There is a general decreasing trend in the number of frost days over the past 54 years in Punjab. Frost days has decreased by 7 days in Islamabad (at 95 % level), by 2 days in Lahore (at 95 % level), by 3 days in Multan (at 95 % level), and by 2 days in Bawalnagar (at 95 % level) in the past 54 years. In Punjab, the total precipitation has a significantly increasing trend in Bahawalnagar (162 mm in 54 years at 95 % level) and in Khanpur (84 mm in 54 years at 95 % level). In Punjab, there is a general increasing trend in the number of days with heavy precipitation. Significant trends are observed in Lahore and Bahawalnagar with 6 days increase in past 54 years at 95 % level of confidence. The number of days with very heavy precipitation also displays a general increasing trend in the past 54 years data analysis. Significant trend is observed in Bahawalnagar with 3 days increase at 95 % level.

Annual total precipitation amount of heavy precipitation events in Punjab show increasing trends in Islamabad, Lahore, Bahawalnagar, and Khanpur, while they show decreasing trends in Sialkot and Multan. Statistically significant increase in the annual total precipitation amount of heavy precipitation events is observed in Khanpur (70 mm in 54 years at 95 % level). Total precipitation amount of very heavy precipitation events in Punjab has displayed increasing trends in Islamabad, Lahore and Bahawalnagar, while it has shown decreasing trends in Sialkot and Multan. Significant trends are observed in the total precipitation amount of very heavy recipitaion events in Islamabad with 112 mm increase and in khanpur with 42 mm increase, both at 95 % level over the past 54 years analysis. 5 day consecutive precipitation amount in Punjab has displayed significant increase in Khanpur with 70 mm increase at 95 % level of confidence. Summer days with maximum temperature > 25°C in Punjab display altogether an increasing trend. Statistically significant trends are observed in Islamabad (20 days increase at 99 % level), in Sialkot (15 days increase at 95 % level), and in Multan (14 days increase at 95 % level) over the past 54 years analysis. Summer days with maximum temperature > 40°C in Lahore displays a significant decreasing trend, while summer days in Bahawalnagar with maximum temperature > 42°C displays a significant increasing trend over the analysis of past 54 years data analysis. The trend in Lahore is 11 days decrease with significance at 95 % level, whereas the trend in Bahawalnagar is 17 days increase with significance at 95 % level.

There is a statistically significant trend observed in the mean of the maximum temperature of Islamabad with increase of 1°C over the past 54 years. Mean of minimum temperature in Punjab is displaying altogether a significant increasing trend in all the observed stations. In Islamabad, it displays a trend of 1.1°C increase, in Sialkot, it displays a trend of 1.8°C increase, in Lahore, it displays a trend of 2.6°C increase, in Multan, it displays a trend of 2°C increase, in Bahawalnagar, it displays a trend of 2.2°C increase, and in Khanpur, it displays a trend of 1.6°C increase, all significant at 95 % level over the past 54 years. Cool nights in Punjab display a significantly decreasing trend altogether with 11 % decrease in Islamabad, 14 % decrease in Sialkot, 23 % decrease in Lahore, 13 % decrease in Multan, 19 % decrease in Bahawalnagar, and 11 % decrease in Khanpur, all confident over 95 % level. Warm nights in Punjab have altogether an increasing trend over the past 54 years analysis. The warm nights has increased in Islamabad by 5 %, in Sialkot by 14 %, in Lahore by 18 %, in Bahawalnagar by 16 %, and in Khanpur by 11 %, all significant at 95 % confidence level. Monthly minimum value of daily minimum temperature display over all an increasing trend over the past 54 years analysis. Significant trends are observed in Lahore (4.3°C increase), in Multan (3.1°C increase), in Bahawalnagar (2.7°C increase), and in Khanpur (2.1°C increase), at 95 % level of confidence. Monthly maximum value of daily minimum temperature in Punjab show variable trends. However significant trends are observed in Lahore (1°C increase), and in Multan (1°C decrease), both with confidence at 95 % level over the past 54 years.

The trends in tropical nights over Punjab display a statistically significant increase in the past 54 years of data. Tropical nights have increased in Islamabad with 22 nights, in Sialkot with 4 nights, in Lahore with 35 nights, in Multan with 18 nights, in Bahawalnagar with 35 nights, and in Khanpur with 23 nights, all significant at 99 % level of confidence. Cool days are decreasing significantly in Islamabad at the rate of 5 days in 54 years significant at 95 % level. Warm days in Islamabad are significantly increasing in Islamabad (6 days in 54 years), and Bahawalnagar (9 days in 54 years) while they are decreasing in Lahore (6 days in 54 years), significant at 95 % level. Monthly minimum value of daily maximum temperature displays a decreasing trend in Punjab with significant decrease in Sialkot (3.1°C), in Lahore (2°C), in Multan (3.8°C), and in Bahawalnagar (5°C), over the last 54 years. Monthly maximum value of daily maximum temperature displays an increasing trend in Punjab. Statistically significant trends are observed in Multan (1.4°C increase), in Bahawalnagar (2°C increase), and in Khanpur (1.2°C increase), all confident at 95 % level. Warm spell duration indicator displays decreasing trends in Sialkot, Lahore, Multan and Khanpur, while it displays increasing trends in Islamabad and Bahawalnagar. However, the trends of both increasing and decreasing warm spell duration are insignificant at both 99 % and 95 % level.

5.4 Observed indices of Sindh

The monthly mean difference between the maximum temperature and the minimum temperature i.e. the diurnal temperature range (DTR) exhibits two distinct trends in the analysis. One of the trends is decreasing with significance at 99% and 95% for Badin and Karachi respectively, that are along the coast (Figure 5.4a and 5.4b). The rate of decrease over the past 54 years (1960-2013) has been estimated to be 0.04°C per year for Badin and 0.01°C for Karachi. On the other hand, an increasing trend of 0.01°C in Chhor, and of 0.03°C in Nawabshah with significance at 91% and 99% respectively is also observed. This signals a significant increase of DTR in the internal Sindh and a significant decrease of the same along the coastal areas of Sindh. Annual total wet-day precipitation in Sindh does not show any significant linear trend over the past 54 years. There is an insignificant increasing trend observed from 1960-1980, then an insignificant decreasing trend from 1981-2000, and then again an insignificant increasing trend onwards till 2013 for Badin, Karachi, and Nawabashah. However, in Chhor, an increasing trend of 2.3 mm/year is observed with confidence of 86% which indicates that Chhor has received an estimated 150 mm more rain over the past 54 years from 1960-2013.

In Sindh, for maximum number of consecutive wet days, there is no significant increasing or decreasing trend observed in Karachi, Chhor, Nawabshah, and Rohri over the past 54 years (1960-2013). However, a weak significant increasing trend at 92% level in Badin suggests an increase of 3 days in the maximum number of consecutive rainy days over the past 54 years. Annual count of heavy precipitation days with precipitation > 10 mm/day has no significant linear trend over Sindh when past 54 years (1960-2013) data is analysed. In Sindh there is an insignificant increasing linear trend in the annual count of heavy precipitation days from 1960 to 1980, then an insigniticant decreasing linear trend from 1981 to 1995, and then once again an insignificant linear increasing trend from 1996 to 2013 and onwards. This indicates no significant change in the heavy precipitation trends over Sindh over the past 54 years. Annual count of days with very heavy precipitation in Sindh also displays an insignificant linear trend with an insignificant increase from 1960 to 1980, an insignificant decrease from 1981-1998, and an insignificant increase 1999 onwards till 2013. This indicates that heavy precipitation days with precipitation > 20 mm/day on annual basis has no significant trend observed in the past 54 years over the Sindh.

Annual total precipitation for very wet days (rainfall rate > 95th percentile) displays variable trends between the stations in Sindh, yet none of them are significant at 95% confidence level. In Badin, Rohri and Chhor, there is an insignificant increasing trend from 1960 to 1990, after which it starts decreasing insignificantly till 2013. Karachi and

Nawabshah also displays insignificant linear trends for the past 54 years. In the recent years, from 1990 to 2013, Chhor has displayed a decrease of 100 mm in annual total of very wet days for the last 24 years. On the other hand Karachi has displayed a simultaneous increase of 100 mm in the annual total precipitation for very wet days over the recent past. Annual total precipitation for extremely wet days over Sindh displays insignificant linear trends over the past 54 years (1960-2013). Insignificant increasing trends are observed in Badin (0.36 mm/year) and in Rohri (0.67mm/year). Nevertheless statistically significant (at 95% level) increasing linear trend is observed in Chhor, which indicates 1.2 mm/year rise in the annual total precipitation amount for extremely wet days over the last 54 years. Analysis of the monthly maximum consecutive 5-day precipitation over Sindh has displayed majorly insignificant trends over the past 54 years. However, for Chhor, it is seen that the index has displayed a statistically significant (at 95% level) increase over the past 54 years. The rate of increase is 1.7 mm/year which approximately amounts to 90 mm increase in the monthly maximum 5-day precipitation over Chhor in the past 54 years. Simple daily intensity index calculated by the annual total precipitation divided by the number of wet days displays an insignificant decreasing linear trend in Badin, whilst it displays an insignificant increasing linear trend in Rohri over the past 54 years over Sindh. Badin displays a straight away decreasing trend with the exception of 1980 to 1995 where it represents that higher amount of precipitation fell over few wet days. In case of Rohri, though the trend is decreasing, yet the 1980 to 1985 high intensity signal is again visible which displays anomalous high precipitation amount in smaller number of days.

Summer days in which maximum temperature is > 25°C is an important index to be analyzed for humid regions like Sindh, since even a degree Celsius rise in humid environment bring distress in the ambient temperature over the region. The region displays statistically significant increase (at 95% level) in Chhor, Karachi, Nawabshah and Rohri, with linear increasing trends of 11 days in Chhor, 26 days in Karachi, 20 days in Nawabshah, and 16 days in Rohri, over the past 54 years. Badin show a decreasing linear trend but is not significant at 95% level. Increase in summer days over Sindh province is statistically significant at 95% level in Chhor, Karachi, and Nawabshah, with linear trend increase of 19 days in Chhor (tmax > 42°C), 19 days in Karachi (tmax > 36°C), and 26 days in Nawabshah (tmax > 44°C). Badin displays a linear decreasing trend of 18 days (tmax > 39°C) which seems significant at 95% level, but not might be so, owing to the missing values between 1965 to 1975, and to the recent increase in the number of summer days after year 2000 till 2013. Rohri displays no significant change in its historical summer day evolution. In Sindh the mean of the maximum temperature over the past 54 years has shown a significant increase of 1.5°C in Chhor, 1.5°C in Karachi, and 1.7°C in Nawabshah over the past 54 years. Badin displays an insignificant linear trend with a decrease from 1960 to 1980, and then a continuous increase afterwards till 2013. Rohri also displays no significant linear trend over the past 54 years.

Increase in minimum temperature renders night time temperature to become warmer. Sindh has displayed significant increase in mean of the minimum temperature in Badin (1.6°C), and in Karachi (2.2°C). Chhor, Nawabshah, and Rohri does not show a significant linear trend, however, in the recent past (2000-2013), the stations display a warming trend in the mean of the minimum temperature. Sindh has displayed a statistically significant decrease in the percentage of cool nights in Badin and Karachi (both stations display 16.2% decrease at 99% level of confidence), and in Chhor (decrease of 6.5% at 86% level of confidence). No statistically significant linear trend is observed in the percentage change of cool nights in Nawabshah and Rohri. In Sindh, the analysis of past data has shown that Badin, Karachi, and Rohri has shown significant increase (at 99% level of confidence) in the number of days with warm nights. The rate of increasing trend in Badin is 10 days, that of Karachi is 16 days, and that of Rohri is 14 days over the past 54 years. There is no statistically significant change in the rate of warm nights of Chhor and Nawabshah, which indicates that the coastal areas (Karachi and Badin) have been more prone to endure increase in the warm nights over the past.

The monthly minimum value of daily minimum temperature in Sindh has displayed significant increasing trend in Badin, Chhor and Karachi with 99% level of confidence in the trend. There is an increase of 2.9°C in the monthly minimum value of daily minimum in Badin, of 2.6°C in the monthly minimum value of daily minimum in Karachi, and of 2.7°C in the monthly minimum value of daily minimum in Chhor. No significant trend is displayed in Nawabshah and Rohri. No significant trend is observed in the monthly maximum value of daily minimum temperature in Sindh. Tropical nights are annual count of days when daily minimum is greater than 20°C. In Sindh, the tropical nights have shown a statistically significant increasing trend at 99% level of confidence in Badin, Chhor, and Karachi. Tropical nights, on average, have been increased by

25 days in Badin, by 14 days in Chhor, and by 36 days in Karachi over the past 54 years. No significant trend is observed Nawabshah and Rohri. Percentage of days when maximum temperature exceeds 90th percentile, is referred to as warm days index. In Sindh, Chhor, Karachi, and Nawabshah display statistically significant increasing trend at 99% level with average increase of 9 days, 8 days, and 11 days respectively. Badin also displays a significant decreasing trend, however, owing the non-availablity of data from 1965-1975, the trend might be insignificant. Rohri shows an increasing trend significant at 80% level of confidence.

Monthly maximum value of daily maximum temperature in Sindh has shown, in general, increasing trend, with Chhor and Nawabshah being statistically significant at 95% and 99% respectively. Chhor displays an increase of 2.2°C (however, recent 15 years from 1998-2013 has shown a decrease), whereas Nawabshah displays an increase of 3.5°C in the monthly maximum value of daily maximum temperature over the past 54 years. In Sindh, Karachi and Nawabshah both display an increasing trend in warm spell duration index at 95% confidence level, while Rohri displays an increasing trend in the index at 83% confidence level. In Karachi, the warm spell duration has increased by six days, in Nawabshah by eight days, and in Rohri by five days over the past 54 years. Badin displays a significant decreasing trend but with smaller confidence owing to the missing values from 1965 to 1975. Chhor displays an increasing trend in the far historical period, yet it gets decreasing in the recent past. It is seen from the analysis that no significant change in the trend of cold spell duration index has occurred over the past 54 years. General trend is increasing but is not statistically significant.

5.5 Observed indices of Baluchistan

Cold spell duration indicator in Baluchistan display significant decreasing trends in Quetta, Dalbandin and Panjgur (Figure 5.5a and 5.5b). The decreasing cold spell in the last 54 years in Quetta is 36 days, in Dalbandin is 17 days, and in Panjgur is 15 days, all significant at 99 % confidence level. No statistically significant trend is observed in the consecutive wet days observed in all stations of Baluchistan. Statistically significant increasing DTR trends are observed in Baluchistan. In Zhob, the increasing trend is 2.2°C (at 95 % level), in Barkhan, the increasing trend is 3.4°C (at 99 % level), and in Dlabandin, the increasing trend is 0.8°C (at 95 % level), in the past 54 years. There are significant decreasing trends in the frost days of Baluchistan. Statistically significant decreasing trends are observed in Quetta (28 days at 95 % level), and in Dalbandin (18 days at 99 % level). Annual total precipitation amount has no significant trend observed in Baluchistan over the past 54 years. No significant trend in heavy and very heavy precipitation amount is observed in Baluchistan. Also no significant trend is observed in the very wet days of Baluchistan in the past 54 years. Moreover, no significant trend is observed in the extremely wet days of Baluchistan in the past 54 years. Furthermore, monthly maximum consecutive 5-day precipitation amount has no significant trend observed in Baluchistan over the last 54 years.

Summer days in Baluchistan display a significant increasing trend upon the analysis of last 54 years. In Zhob, there is a 26 days increase, in Khuzdar, there is a 36 days increase, in Quetta, there is a 29 days increase, in Barkhan, there is a 32 days increase, in Dalbandin, there is a 32 days increase, and in Panjgur, there is a 23 days increase, in the summer days over the last 54 years, all significant at 99 % level of confidence. Extreme summer days in Baluchistan also have a significant increasing trend. Summer days with maximum temperature > 36°C displays an increasing trend of 23 days in Quetta, those with maximum temperature > 43°C displays an increasing trend of 28 days in Dalbandin, and those with maximum temperature greater than 40°C displays an increasing trend of 16 days, over the last 54 years analysis. There is a significant increasing trend in the mean of the maximum temperature observed in Baluchistan. In Zhob, the increasing trend is of 1.4°C, in Khudar, the increasing trend is of 1.8°C, in Quetta, the increasing trend is of 2.1°C, in Barkhan, the increasing trend is of 1.4°C, in Dalbandin, the increasing trend is of 2.7°C, and in Panjgur, the increasing trend is of 1.4°C, over the last 54 years, all significant at 95 % level of confidence. Stations of Baluchistan display significant increase, except for Zhob and Badin (which show insignificant decrease) in the trends of mean of the minimum temperature index. Statistically significant trends of the index are observed in Quetta with 2.8°C increase, in Dalbandin with 1.6°C increase, and in Panjgur, with 1.4°C increase, over the last 54 years.

Cool nights display a significant decreasing trend in Quetta, Dalbandin, and Panjgur stations of Baluchistan. In Quetta, the index is decreased by 23 nights, in Dalbandin, the index is decreased by 10 nights, and in Panjgur, the index is decreased by 11 nights, over the past 54 years. Warm nights in Baluchistan display a significant increasing trend in Khuzdar, Quetta, and Panjgur stations. In Khuzdar, the index trend displays an increase of 15 nights, in Quetta, the index trend displays an increase of 9 nights, and in Panjgur, the index trend displays an increase of 6 nights, in the last 54 years. There is no significant trend observed in the monthly minimum value of daily minimum temperature in Baluchistan over the last 54 years. No significant trends are observed in the monthly maximum of minimum temperatures over Baluchistan in the past 54 years. However, Zhob displays a significant trend in the index with a decrease of 2.9°C at 95 % confidence level. Tropical nights display a significant increasing trend in Quetta (27 nights at 99 % level), in Dalbandin (28 nights at 99 % level), and in Panjgur (31 nights at 99 % level), over the past 54 years analysis.

Cool days are significantly decreasing all over Baluchistan. Zhob displays a decrease of 5 days, Khuzdar displays a decrease of 10 days, Quetta displays a decrease of 11 days, Dalbandin displays a decrease of 13 days, and Panjgur diplays a decrease of 7 days, significant at 95 % level, over the last 54 years. There is a statistically significant increase in the number of warm days in Baluchistan over the last 54 years data analysis. Increasing trends of the index are displayed in Zhob (8 days at 99 % level), in Khuzdar (10 days at 95 % level), in Quetta (12 days at 99 % level), Barkhan (10 days at 95 % level), in Dalbandin (16 days at 99 % level), and in Panjgur (9 days at 99 % level), over the last 54 years. No significant trends are observed in monthly minimum of the daily maximum temperature over Baluchistan in the last 54 years. Significant trends are observed in the monthly maximum of the daily maximum temperature over Baluchistan. Quetta displays 1.5°C increase in the index, and Dalbandin displays a 2.4°C rise in the index, both significant a t 99 % level, over the past 54 years. Warm spell duration indicator displays increasing trends over Baluchistan. The index displays statistically significant trends in Quetta (8 days increase), in Barkhan (16 days increase), in Dalbandin (18 days increase), and in Panjgur (10 days increase), all at 95 % level of confidence.

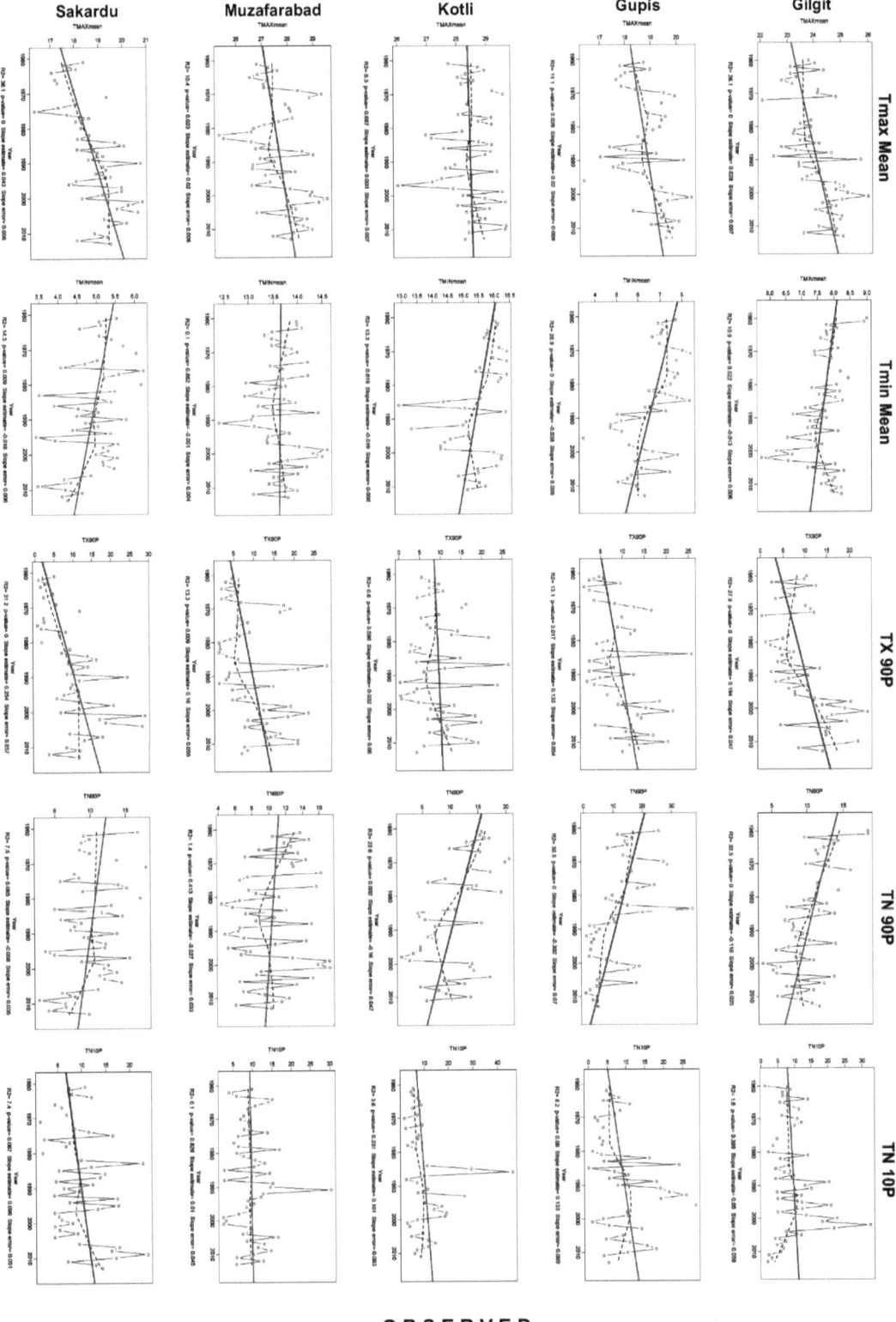

Figure 5.1: Significant observed climate indices over GB-AJK region.

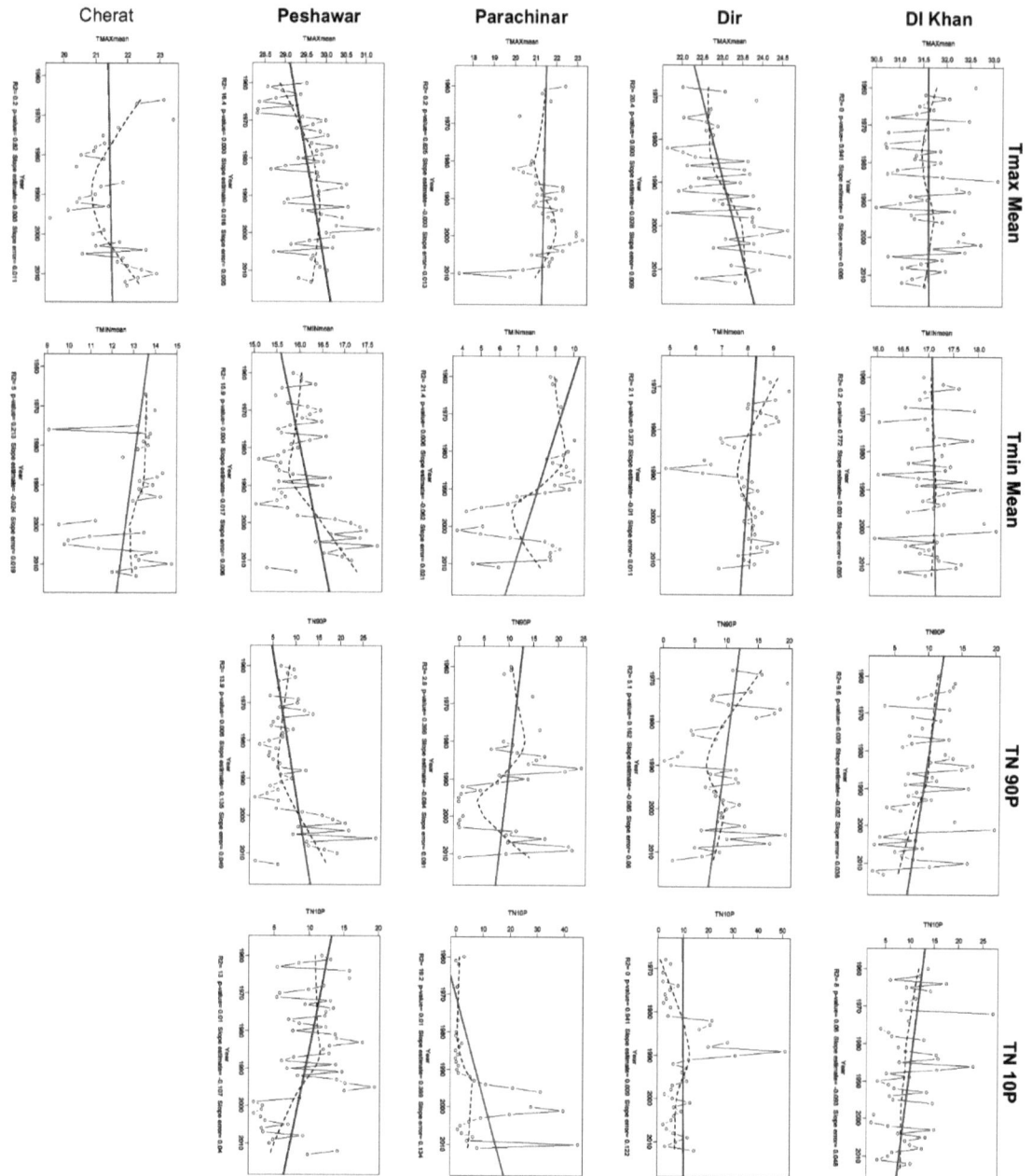

Figure 5.2: Significant observed climate indices over KPK province.

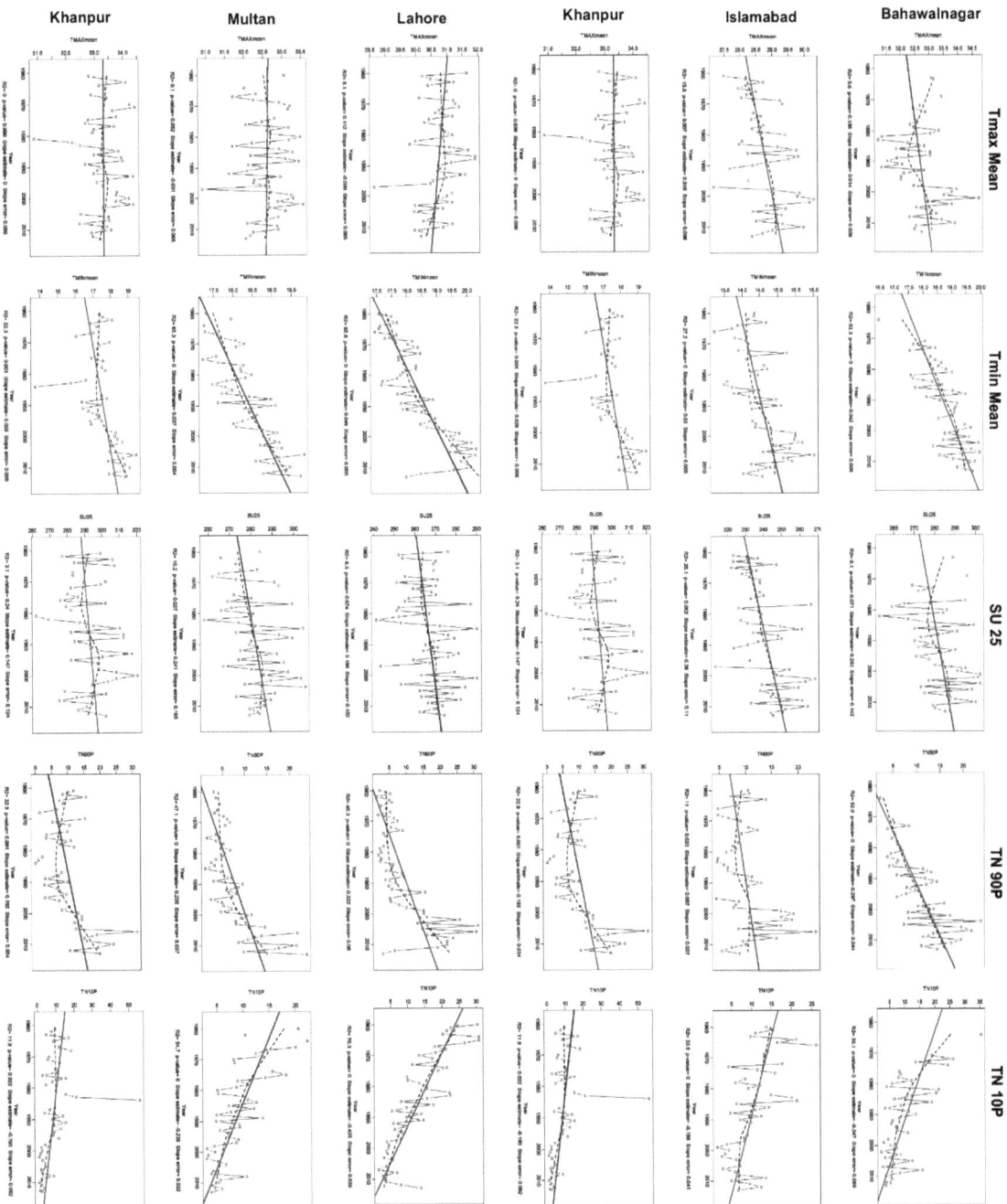

Figure 5.3: Significant observed climate indices over Punjab province.

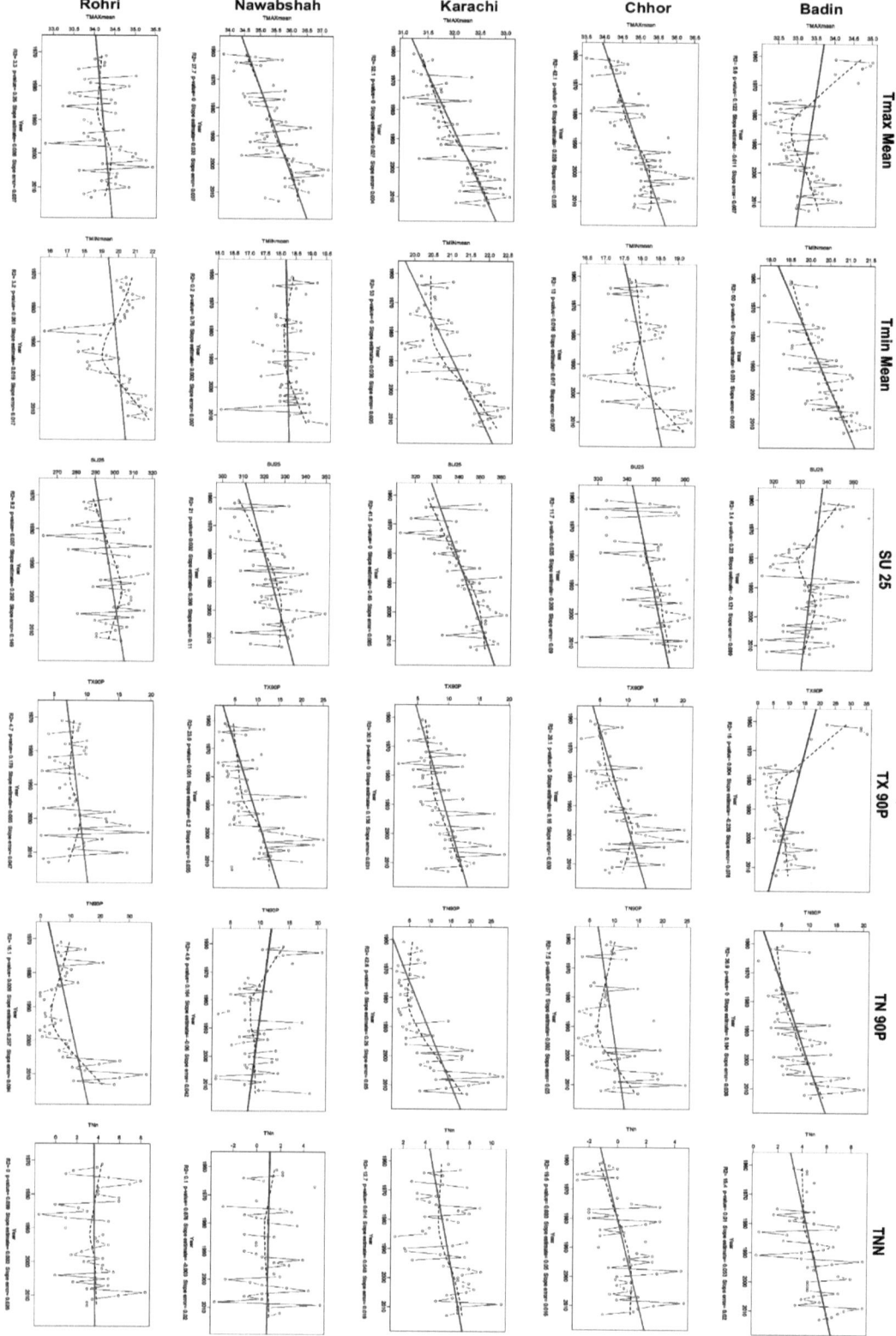

Figure 5.4a: Significant observed climate indices over Sindh province.

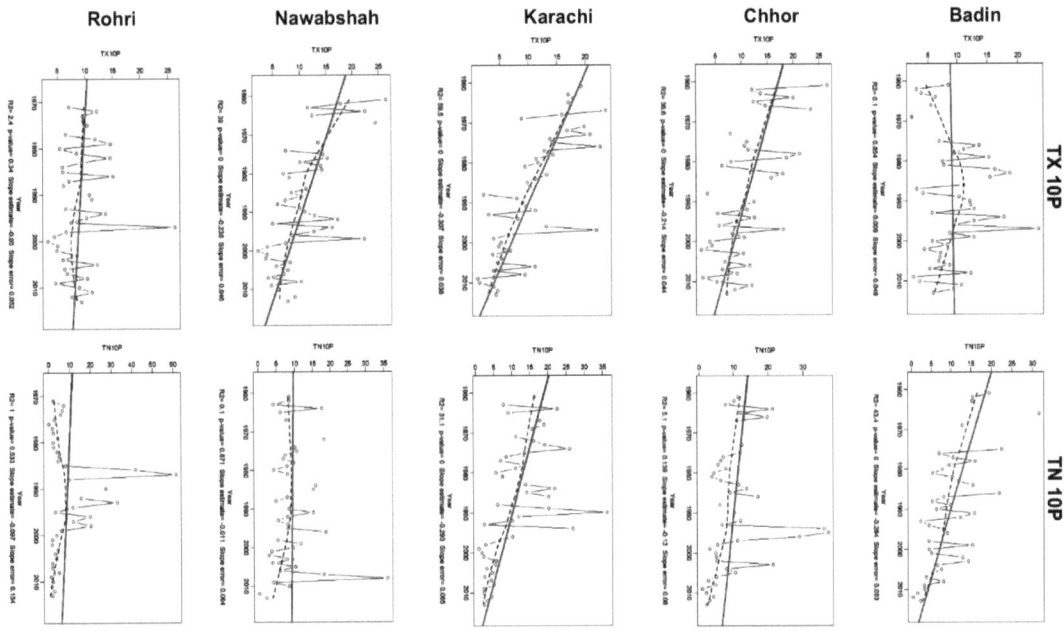

Figure 5.4b: Significant observed climate indices over Sindh province (Continued)

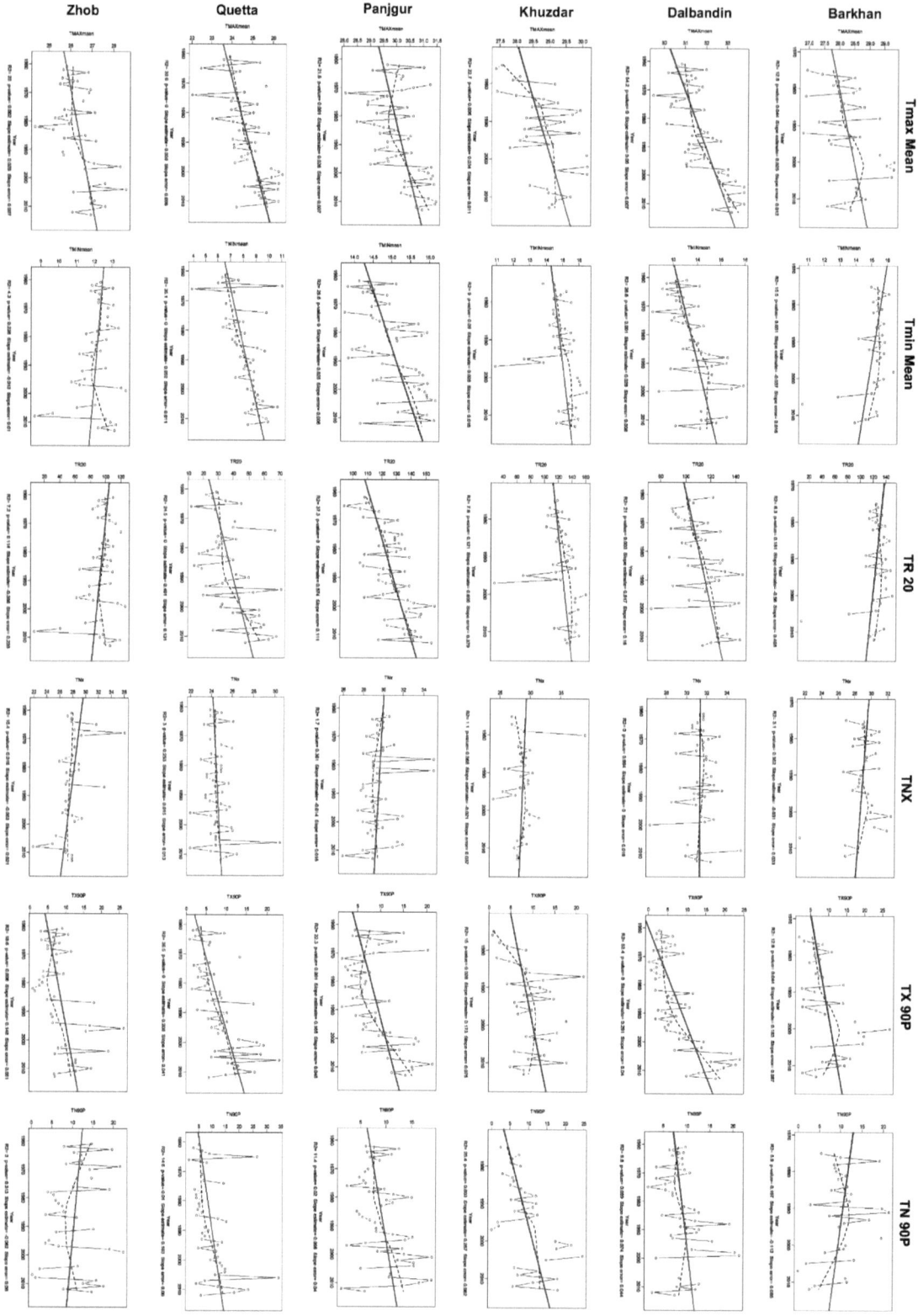

Figure 5.5a: Significant observed climate indices over Baluchistan province.

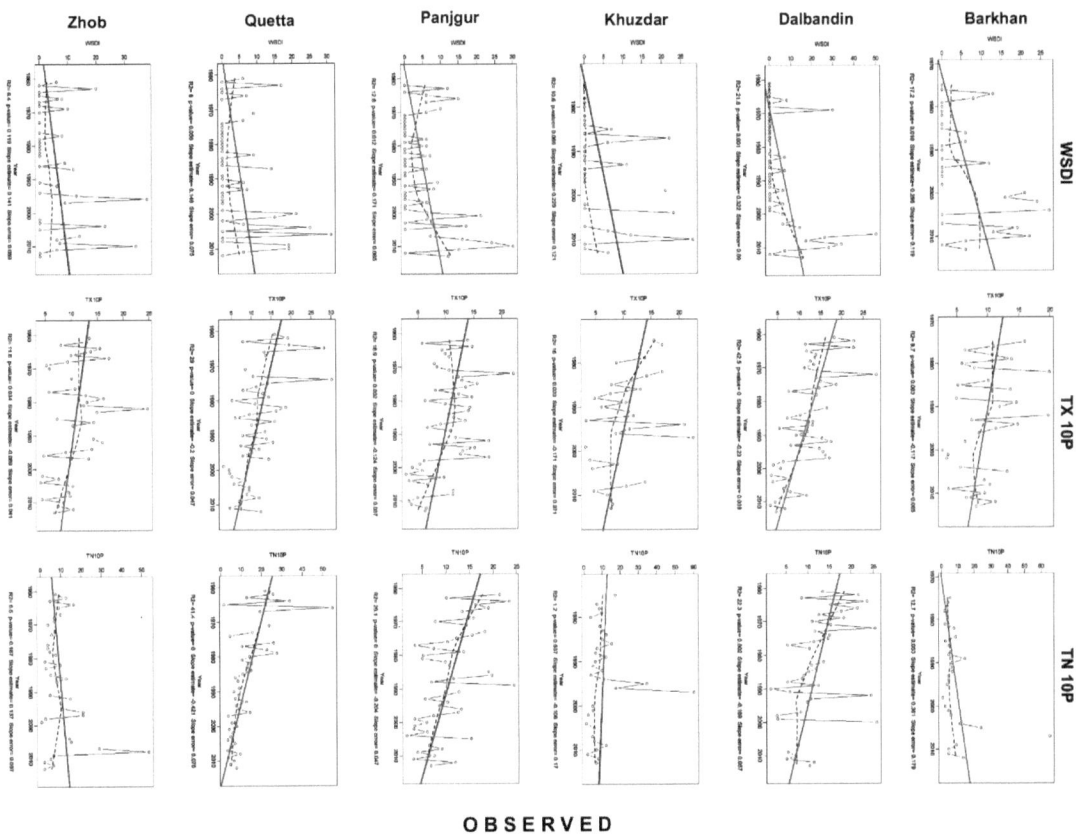

Figure 5.5b: Significant observed climate indices over Baluchistan province (continued).

5.6 Observed Indices of All Pakistan

Both mean maximum and mean minimum temperatures are increasing significantly by 0.8°C and o.9°C respectively over whole Pakistan (Figure 5.6). Summer days are shown to increase significantly by 15 days, while tropical nights are shown to increase significantly by 18 nights over whole Pakistan. Monthly minimum of daily maximum temperature has increased by 1.5°C, monthly minimum of daily minimum temperature has increased by 1.8°C, whereas, monthly maximum of daily minimum temperature has significantly decreased by 0.8°C. Frost days are decreasing significantly by 5 days in 54 years. Both cool days and cool nights are decreasing significantly by 7 days and 14 nights respectively over whole Pakistan. It is also seen that the cold spell duration index has decreased by 29 days over the 54-year period in whole Pakistan (see Figure 5.7 for spatial trends).

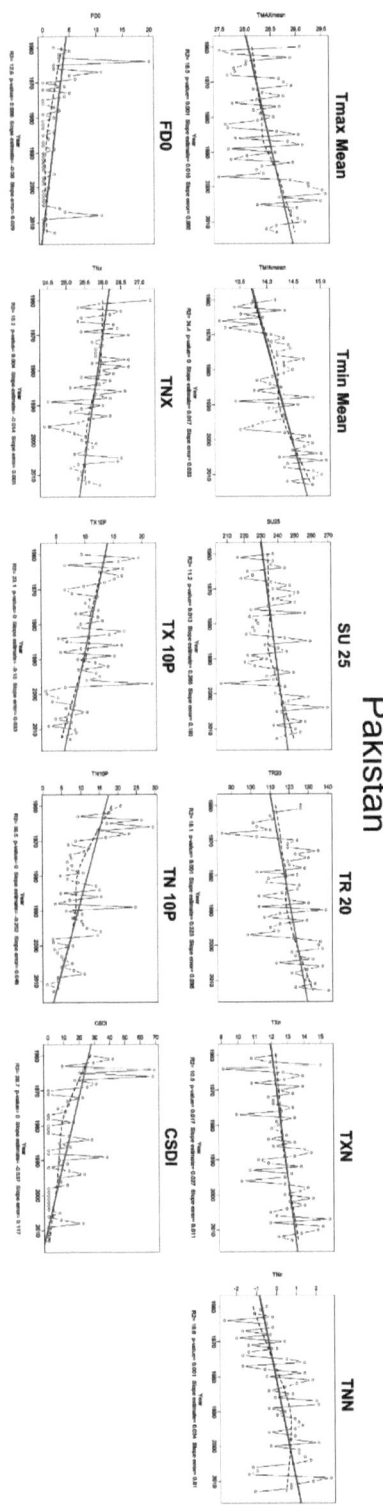

Figure 5.6: Significant observed climate indices over whole Pakistan.

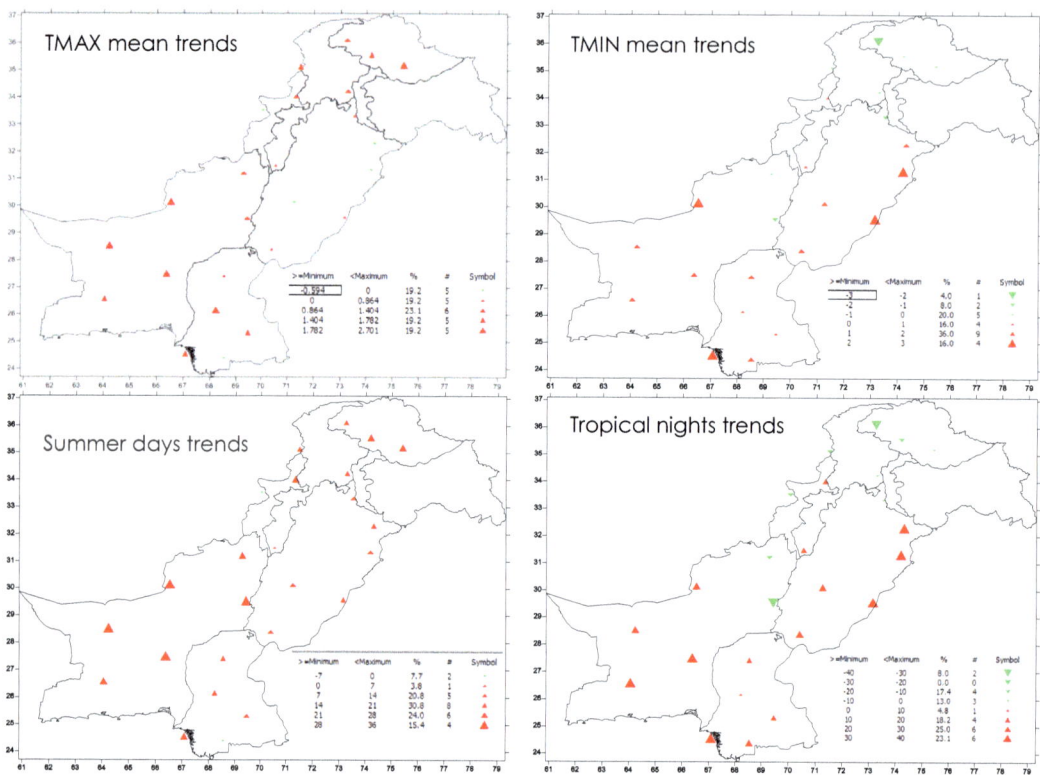

Figure 5.7: Spatial historical (1960-2013) climate trends over Pakistan

6 GCMs AND RCMs SIMULATED INDICES

Statistically significant increasing and decreasing trends are displayed by GCM ensembles. Taking the ensemble mean of the GCMS emulating observed climatology gives more confidence over the generated results. Province based analysis of the indices is done over a historical period of 35 years (1970-2004) and the results discussed.

6.1 GCM ENSEMBLES SIMULATED INDICES

6.1.1 GCM ensembles simulated indices over GB-AJK

Mean maximum temperatures in GB-AJK are seen to simulate significant increasing trend. The rate of increase is 0.03°C/year as analyzed by GCMs simulated data. The results conform to the observed increase in the maximum temperatures over the region. Mean minimum temperatures in GB-AJK display an increasing trend of 0.02°C/year as analyzed over the GCMs simulated data. Summer days in GB-AJK have increased by 10 days in the GCMs simulated data. Observed data analysis has also shown an increasing trend in the summer days index over GB-AJK. Ice days have significantly decreased with a trend of 19 days. The decrease in ice days is also evident in the observed data, especially over the GB region. Monthly maximum value of daily maximum temperature in GB-AJK displays an increase of 0.03°C as analyzed over the historical simulated GCMs ensemble data. Monthly minimum value of daily maximum temperature displays an even higher trend (0.07°C) which is also a feature indicated by observed historical data. Monthly maximum value of daily minimum temperature displays an increase of 0.02°C while monthly minimum value of daily minimum temperature displays an increase of 0.08°C in GB-AJK region. Cool days are seen to display a significant decreasing trend of 11 days decrease while, on the other hand, warm days are seen to display 13 days increase as simulated by GCMs historical data. Cool nights display a decreasing trend of 11 nights, while warm nights display an increasing trend of 10 days in GCMs simulated historical ensemble data.

6.1.2 GCM ensembles simulated indices over KPK

In KPK, mean maximum temperature displays an increasing trend of 0.02°C/year in the GCMs simulated historical ensemble data. Mean minimum temperature is also seen to display an increasing trend of 0.03°C/year over the KPK province. Monthly minimum value of daily maximum temperatures display a significant increase of 0.08°C in the GCMS simulated historical ensemble data. Monthly maximum value of daily minimum temperature also displays an increase of 0.02°C over the KPK province. Moreover, monthly minimum value of daily minimum temperature is seen to display an increasing trend of 0.09°C over the KPK province. Cool days are decreasing significantly at the rate of 7 days/35 years, while cool nights are decreasing significantly at the rate of 10 days/ 35 years as displayed by the GCMs historical ensemble data over the KPK province. Furthermore, warm nights display an increasing trend of 9 days as simulated by GCMs historical ensemble data. This conforms to a significant rate of decline in cold spells (12 days decrease per 35 years) over the KPK province.

6.1.3 GCM ensembles simulated indices over Punjab

Both mean maximum and mean minimum temperatures display increasing trends of 0.02°C/year over the Punjab province. Summer days are seen to display a significant increasing trend of 11 days over the past 35 years. Frost days are significantly decreasing (5 days/ 35 years) which is also a significant feature of the observed frost days index. Monthly minimum value of daily maximum temperature is significantly increasing (2.3°C) in Punjab. Moreover, significant increase in monthly minimum value of daily minimum temperature (2.0°C) is seen in the GCMs simulated historical ensemble data. Cool days are decreasing significantly at the rate of 6 days/ 35 years, while cool nights are decreasing significantly at the rate of 9 nights/ 35 years over the Punjab province. Warm nights are displaying a significant increase of 9 nights/ 35 years as simulated by GCM historical ensemble data over Punjab province. In terms of precipitation indices, maximum number of consecutive wet days has increased significantly by 11 days over the Punjab province. The index is also seen to display a significant increasing trend in the observed data of Islamabad city in the Punjab province.

6.1.4 GCM ensembles simulated indices over Sindh

An increasing trend of 0.02°C/year is seen in the mean maximum and the mean minimum temperatures of the Sindh province. Summer days have significantly increased by 18 days in the historical 35 years GCM ensemble data. Monthly minimum value of daily maximum temperature displays a significant increase of 2.1°C, monthly maximum value of daily minimum temperature displays a significant increase of 0.9°C, and monthly minimum value of daily minimum temperature displays a significant increase of 1.8°C over the 35 years historical simulated GCMs ensemble data. Cool days display a decreasing trend (8 days decrease) which conforms to the observed significant decrease in cool days over Sindh province. As per ensemble GCMs results, warm days are increasing simultaneously at the rate of 6 days/35 years over the Sindh province. Cool nights display a decreasing trend of 11 days/35 years, whilst warm nights display an increasing trend of 11 days/ 35 years, as shown by GCMs historical ensemble data. The cold spell duration index is seen to display a significant decreasing trend of 15 days/ 35 years, which is a coherent feature displayed by the observed index over the Sindh province.

6.1.5 GCM ensembles simulated indices over Baluchistan

Mean maximum and mean minimum temperatures, both display significant increasing trends of 0.02°C, as seen through historical GCMs ensemble data over the Baluchistan province. The tropical nights with minimum temperatures exceeding 20°C Celsius are seen to display a significant increasing trend of 12 nights/35 years over the Baluchistan province. The monthly maximum value of daily maximum temperature displays a significant increase of 1.0°C, the monthly minimum value of daily maximum temperature displays a significant increase of 2.4°C, and the monthly maximum value of daily minimum temperature displays a significant increase of 0.7°C, as seen through historical simulated GCMs ensemble data over Baluchistan province. Cool days are seen to display significant decrease of 7 days/ 35 years, whilst warm days are seen to display significant increase of 6 days/35 years. Cool nights are seen to display significant decrease of 9 nights/35 years, whereas warm nights are seen to display significant increasing trend of 11 nights/35 years, as seen through historical simulated GCMs ensemble data over Baluchistan province.

6.2 Statistically downscaled GCMs (Nex-NASA) based climate indices

Province based indices are found using statistically downscaled Nex-NASA ensemble GCMs. Linear trends with their statistical significance are computed and the results discussed.

6.2.1 Nex-NASA based climate indices over GB-AJK

Mean minimum temperature increase is seen to display a significant trend with a magnitude of 0.7°C/year rise in the GB-AJK region. This increase is followed by a significant increase in the monthly maximum value of daily minimum temperature where its magnitude has increased by 0.8°C over the historical period. Warm days are seen to increase significantly at the rate of 9 days/ 35 years over the GB-AJK region. Cool nights display a significant decreasing trend of 8 nights/ 35 years, while warm nights display a significant increasing trend of 13 nights/ 35 years over the GB-AJK region. Warm spell duration index also displays a significant increasing trend of 14 days over the historical period in the GB-AJK region. The index has also shown significant increasing trend over the Skardu city of GB-AJK region upon the analysis of past data. This indicates Skardu in GB-AJK as a vulnerable zone that may be affected significantly by heat waves over the GB-AJK region. The diurnal temperature range is seen to decrease significantly by 0.3°C over the historical period in the GB-AJK region.

6.2.2 Nex-NASA based climate indices over KPK

Both mean maximum and mean minimum temperatures are seen to display significant increasing trends of 0.4°C and of 0.6°C respectively over the historical period in KPK region. The summer days are also seen to display a significant increasing trend of 11 days over a period of 35 years based on historical years of KPK region. Tropical nights are significantly increasing (4 nights/ 35 years) as seen through Nex-NASA historical climate over the KPK province. Warm days are seen to display a significant increasing trend of 7 days/ 35 years, whilst warm nights are seen to display a significant increasing trend of 12 nights/ 35 years over the KPK province. Moreover, cool nights are decreasing significantly at the rate of 8 nights/ 35 years, as seen through Nex-NASA historical climate over the KPK province. Cold spell duration index displays a significant decrease of 12 days over a period of the 35 years historical period. Furthermore, diurnal temperature range is seen to decrease significantly (0.2°C) over the historical period in the KPK region.

6.2.3 Nex-NASA based climate indices over Punjab

Mean maximum and mean minimum temperatures, as displayed by Nex-NASA datasets, suggest significant increases of 0.5°C and of 0.6°C, respectively, over the past climate in Punjab province. Summer days are also seen to increase significantly at a rate of 9 days/35 years over the region. Cool days are significantly decreasing at a rate of 7 days/ 35 years, while warm days are significantly increasing at a rate of 7 days/35 years, as suggested by Nex-NASA output. Cool nights are significantly decreasing at a rate of 9 nights/35 years, whereas warm nights are significantly increasing at a rate of 12 nights/35 years, as displayed by Nex-NASA over the Punjab province. Moreover, it is seen that cold spell duration index has significantly decreased by 18 days over a period of 35 years in the Punjab province.

6.2.4 Nex-NASA based climate indices over Sindh

Both mean maximum and mean minimum temperatures are seen to increase significantly by 0.6°C and by 0.7°C respectively, over the historical period in the Sindh province. Monthly minimum value of daily minimum temperature is seen to increase significantly by 1.0°C, over the 35 year historical period. Cool days are seen to decrease significantly by 9 days/35 years, whilst warm days are seen to increase significantly by 12 days/35 years, over the Sindh province. Cool nights are seen to decrease significantly by 14 nights/35 years, whereas warm nights are seen to increase significantly by 14 nights, over the Sindh province. Moreover, warm spell duration index is seen to display an alarming and significant increasing trend of 14 days over a period of 35 years in the Sindh province.

6.2.5 Nex-NASA based climate indices over Baluchistan

Significant increases in mean maximum and mean minimum temperatures are seen with magnitudes of 0.5°C and 0.7°C respectively, over the Baluchistan province. Tropical nights are also seen to increase significantly by 11 nights over the 35 year historical period. Monthly maximum value of daily minimum temperature is seen to display a significant increase of 0.5°C over the 35 year historical period. Cool days are seen to display a significant decrease of 5 days/35 years, whereas warm days are seen to display a significant increase of 12 days/35 years, over the Baluchistan province. Moreover, cool nights are seen to display significant decrease of 10 nights/35 years, whilst warm nights are seen to display significant increase of 15 nights/35 years, over the Baluchistan province. Furthermore, Nex-NASA also suggests a significant increase of 17 days in the warm spell duration index, over the 35 years historical period in Baluchistan province.

6.3 Dynamically downscaled (CORDEX) based climate indices

The CORDEX based indices are calculated on provincial basis after taking the ensembles where deemed fit based on the discussion done in the "Data and methodology" section. The resulting indices output is robust and may be considered for implications resulting from climate change in the region.

6.3.1 CORDEX based climate indices over GB-AJK

CORDEX suggests a 0.9°C mean minimum temperature increase over the historical 35 years period in the GB-AJK region (Figure 6.1). Cool nights are seen to display a significant decrease of 10 nights, while warm nights are seen to display a significant increase of 8 nights over the historical 35 years period in the GB-AJK region. Moreover, warm spell duration index is seen to increase significantly by 17 days/35 years, whilst simultaneously, cold spell duration index is seen to decease significantly by 17 days/35 years over the 35 years historical period in the GB-AJK region.

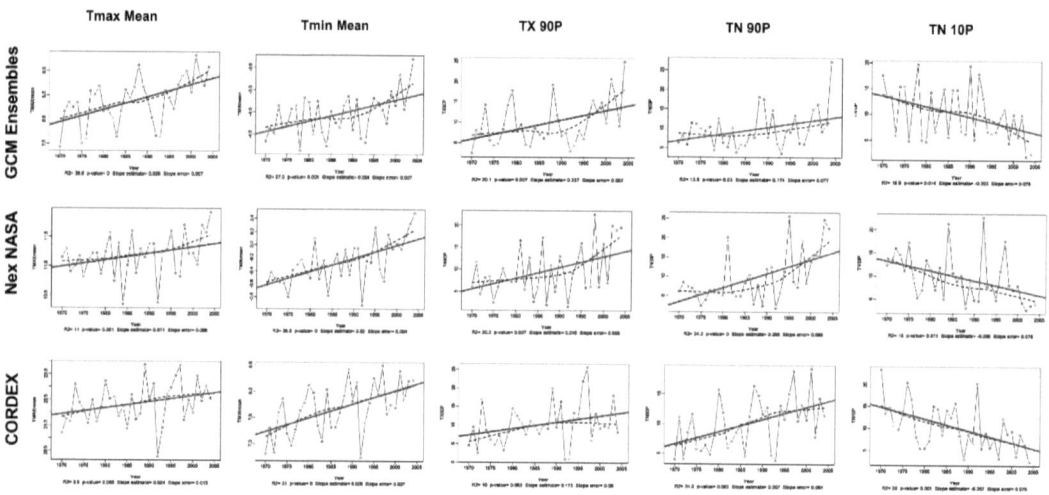

Figure 6.1: Significant simulated, statistically downscaled, and dynamically downscaled climate indices over GB-AJK region.

6.3.2 CORDEX based climate indices over KPK

Both mean maximum and mean minimum temperatures are seen to increase significantly by 1.0°C and by 0.9°C over the KPK region (Figure 6.2). Summer days are seen to increase significantly at a rate of 13 days/35 years in the KPK province. Tropical nights are seen to increase significantly at a rate of 25 nights/ 35 years, whilst frost days are seen to decrease significantly at a rate of 13 days/ 35 years, in the KPK province. Monthly minimum value of daily maximum temperature is seen to increase significantly by 2.2°C over the historical 35 years period in the KPK province. Cool days have significantly decreased by 10 days, whilst warm days have significantly increased by 13 days, as suggested by the CORDEX output over the historical 35 years period in the KPK province. Cool nights have significantly decreased by 8 nights, whereas warm nights have significantly increased by 9 nights, as suggested by the CORDEX output over the historical 35 years period in the KPK province. Moreover, warm spell duration index is seen to display significant increase of 16 days over the past 35 years in the KPK province.

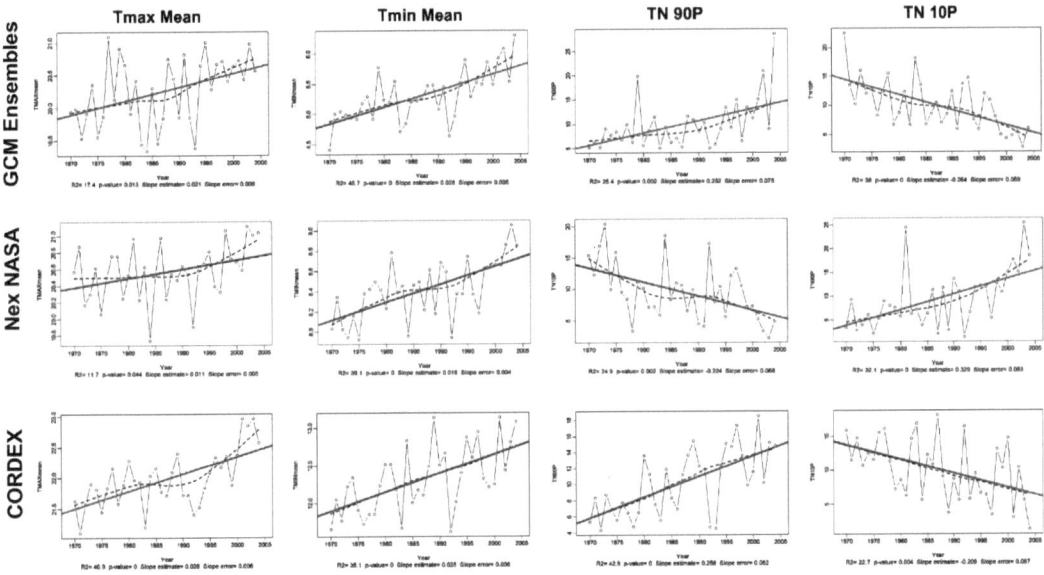

Figure 6.2: Significant simulated, statistically downscaled, and dynamically downscaled climate indices over KPK province.

6.3.3 CORDEX based climate indices over Punjab

CORDEX displays significant increases in both mean maximum and mean minimum temperatures with magnitude 1.0°C, over the Punjab province (Figure 6.3). Summer days are also increasing at a rate of 17 days/35 years, as suggested by CORDEX output over the Punjab province. Monthly minimum value of daily maximum temperature is seen to increase significantly by 1.9°C, whereas monthly minimum value of daily minimum temperature is seen to increase significantly by 1.2°C, over the historical 35 years period in the Punjab province. Cool days are significantly decreasing at a rate of 10 days/35 years, while warm days are significantly increasing at a rate of 11 days/35 years, as suggested by the CORDEX output over the Punjab province. Moreover, it is seen that cool nights are significantly decreasing at a rate of 11 nights/35 years, whilst warm nights are significantly increasing at a rate of 16 nights/35 years, over the Punjab province. Furthermore, annual total precipitation of very wet days is seen to increase significantly by 24 mm over the historical 35 years period in the Punjab province.

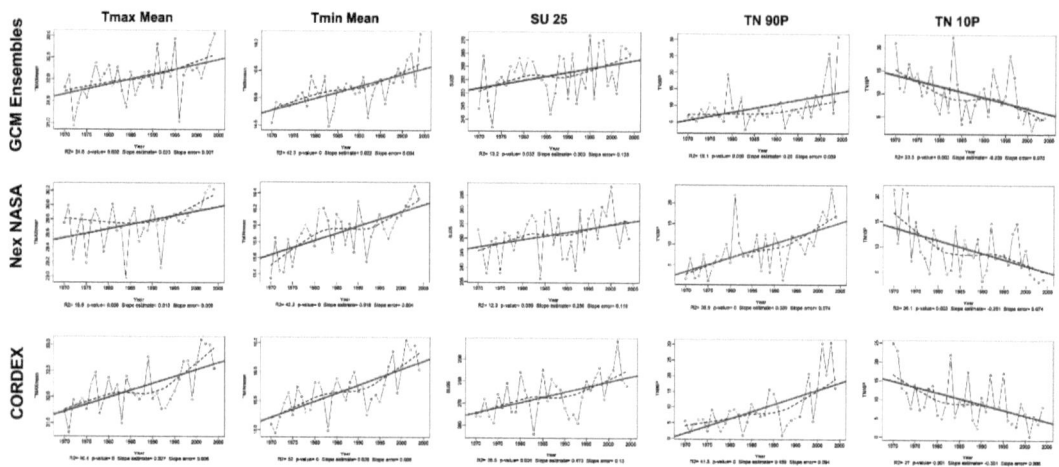

Figure 6.3: Significant simulated, statistically downscaled, and dynamically downscaled climate indices over Punjab province.

6.3.4 CORDEX based climate indices over Sindh

Both mean maximum and mean minimum temperatures display significant increase of 1.0°C over the 35 year historical period in the Sindh province (Figure 6.4). Summer days are also increasing at a rate of 19 days/35 years, based on CORDEX historical climate over Sindh province. Tropical nights are increasing at a rate of 15 nights/35 years, as suggested by CORDEX historical climate over Sindh province. It is seen that monthly maximum value of daily maximum temperature has increased significantly by 1.2°C, monthly minimum value of daily maximum temperature has increased significantly by 1.7°C, monthly maximum value of daily minimum temperature has increased significantly by 0.5°C, and monthly minimum value of daily minimum temperature has increased significantly by 1.8°C, over the 35 years' historical period in Sindh province. Cool days are seen to decrease significantly by 12 days, whilst warm days are seen to increase significantly by 13 days over the historical period of 35 years. Cool nights are decreasing significantly at a rate of 15 nights/35 years, while warm nights are increasing significantly at a rate of 21 nights/35 years, as displayed by the CORDEX in the Sindh province. Moreover, significant increase of 18 days in warm spell duration index, as well as significant decrease of 19 days in cold spell duration index, over the 35 years historical period suggests recurrence of heat waves, which may cause ambient distress among dwellers of the Sindh province.

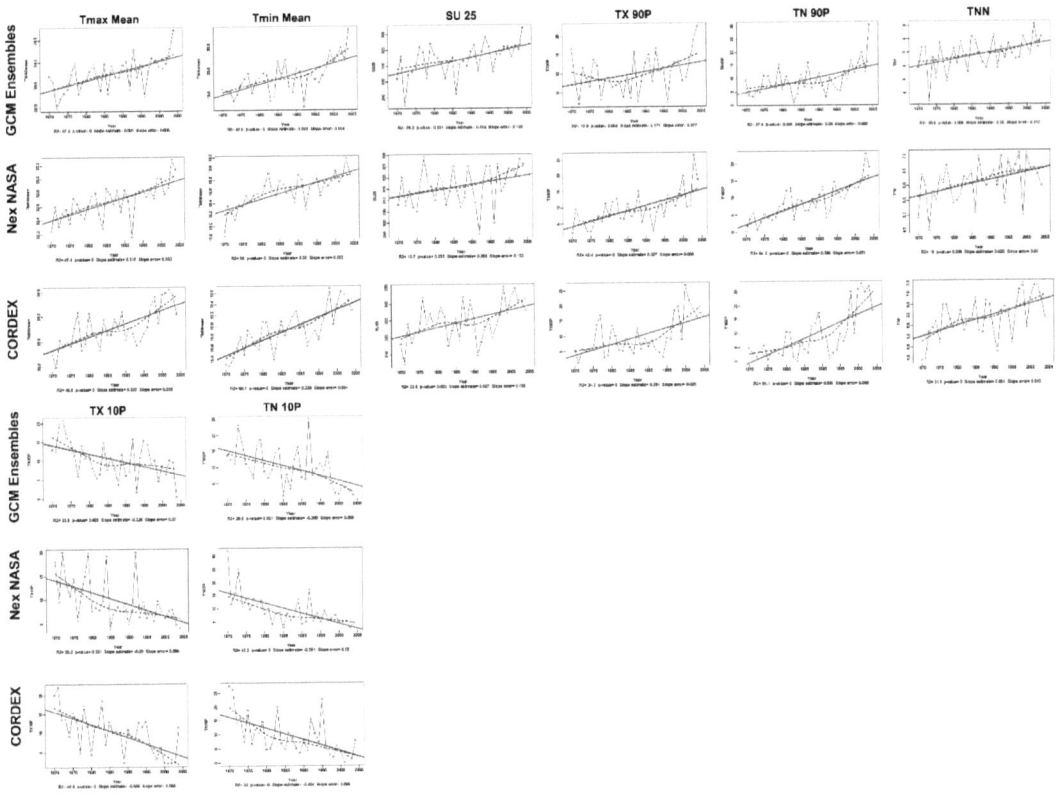

Figure 6.4: Significant simulated, statistically downscaled, and dynamically downscaled climate indices over Sindh province.

6.3.5 CORDEX based climate indices over Baluchistan

Significant increases in mean maximum and mean minimum temperatures are seen with magnitudes of 1.1°C and 1.0°C respectively, over the 35 years historical period in the Baluchistan province (Figure 6.5). Summer days are seen to increase significantly at a rate of 14 days/35 years while tropical nights are seen to increase significantly at a rate of 15 nights/35 years, as displayed by CORDEX output over Baluchistan province. Monthly maximum value of daily minimum temperature has increased significantly by 0.6°C over the historical period, as suggested by CORDEX output over Baluchistan province. Cool days have significantly decreased by 13 days, while warm days have significantly increased by 14 days, as seen by CORDEX output in Baluchistan province. Cool nights are significantly decreasing at a rate of 14 nights/35 years, while warm nights are significantly increasing at a rate of 19 nights/35 years, as seen through CORDEX output over Baluchistan province. Moreover, the CORDEX output displays a significant increasing trend of 21 days/35 years, as well as a significant decreasing trend of 22 days/35 years, over the Baluchistan province.

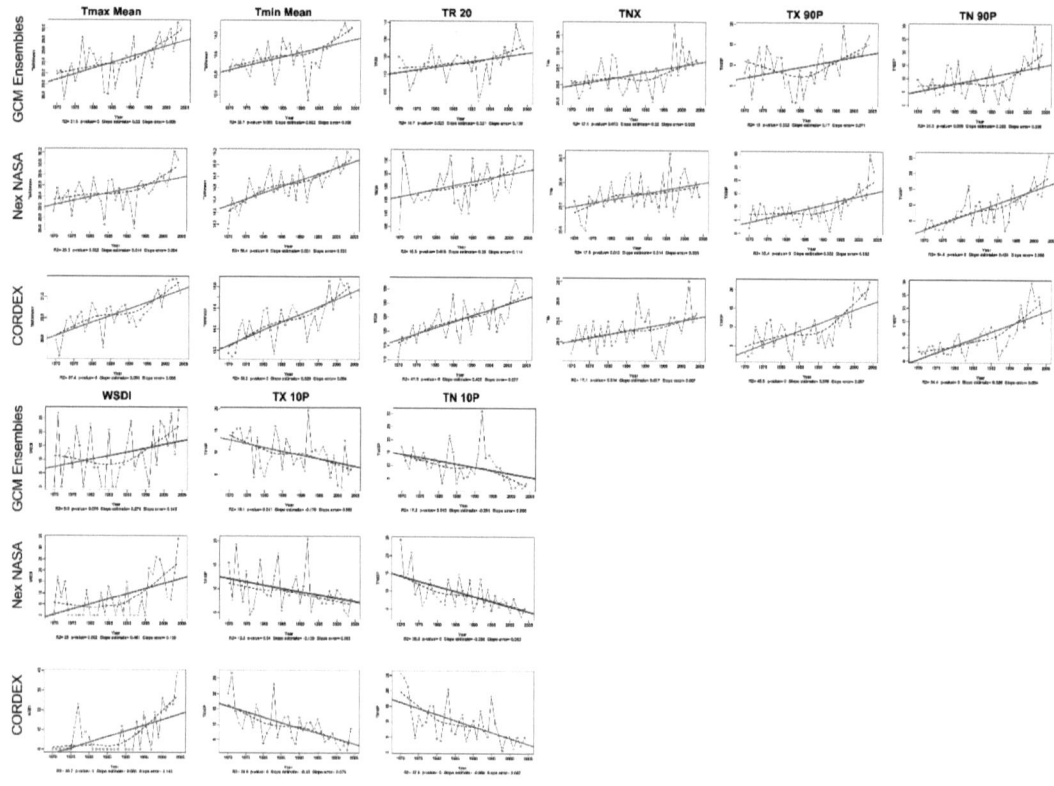

Figure 6.5: Significant simulated, statistically downscaled, and dynamically downscaled climate indices over Baluchistan province.

7 PROJECTED CLIMATE AND INDICES OVER THE PROVINCES OF PAKISTAN

7.1 NEXT 30 YEAR'S ENSEMBLE PROJECTED INDICES

Province based analysis for 2016-2045 projection period is done to assess trends of significant climate indices in the near future. Ensemble mean of the Nex-NASA GCMs is taken for the analysis, owing to its better emulation of both past climatology and Taylor statistics over all the provinces of Pakistan. Two emission scenarios (RCp4.5 and RCP8.5) are taken to assess both low end and high end pathways over the provinces.

7.1.1 Projected indices over Punjab

7.1.1.1 RCP 4.5

Under RCP4.5 protocols, both mean maximum and mean minimum temperatures are projected to increase significantly by 1.0°C in the next 30 years over the Punjab province. Tropical nights are also projected to increase by 14 nights in the next 30 years over the Punjab province. Monthly minimum of daily maximum temperature is projected to increase by 1.8°C, while monthly maximum of daily minimum temperature is projected to increase by 0.8°C over the Punjab province. Monthly minimum of daily minimum temperature is also projected to increase significantly by 1.5°C in the next 30 years over the Punjab province. Warm days are projected to increase significantly by 15 days, whilst warm nights are projected to increase significantly by 19 nights in the next 30 years projection period over the Punjab province. Warm spells are projected to increase significantly by 22 days, while colds spells are projected to decrease significantly by 31 days in the next 30 years over the Punjab province. Moreover, significant decreases in the trends of cold days and cold nights are also projected over the Punjab province in the next 30 years projection period.

7.1.1.2 RCP 8.5

Under RCP8.5 protocols, both mean maximum and mean minimum temperatures are projected to increase significantly by 1.1°C and 1.4°C respectively in the next 30 years over the Punjab province (Figure 7.1). Summer days are projected to increase by 21 days, whilst tropical nights are projected to increase by 9 nights in the next 30 years. Monthly maximum of daily maximum temperature is projected to increase significantly by 1°C, whereas, monthly maximum of daily minimum temperature is projected to increase significantly by 1.7°C over the Punjab province. Warm days are projected to increase significantly by 21 days, whilst warm nights are projected to increase significantly by 25 nights in the projection period. Moreover, cool days are projected to decrease significantly by 13 days, while cool nights are projected to decrease significantly by 26 nights in the next 30 years. Furthermore, ensemble projections show a significant increase of 30 days in the warm spell duration index, as well as a significant decrease of 24 days in the cold spell duration index in the next 30 years projection period over the Punjab province.

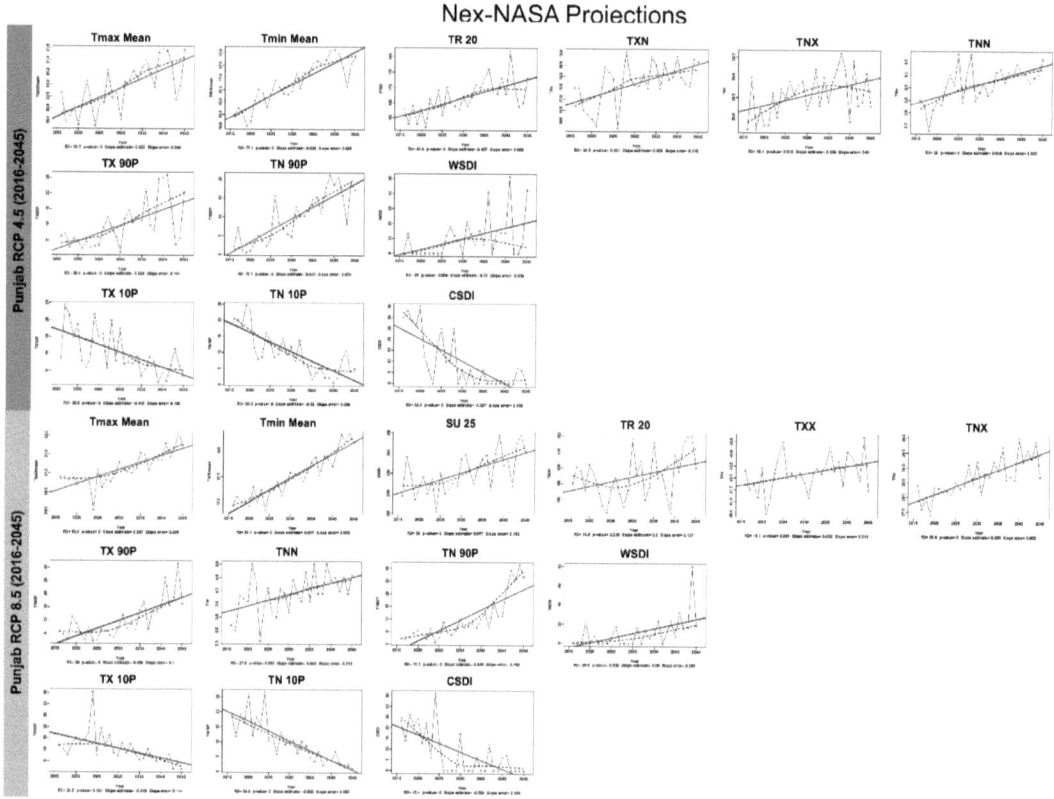

Figure 7.1: Next 30 year's (2016-2045) ensemble projected indices over Punjab province.

7.1.2 Projected indices over Baluchistan

7.1.2.1 RCP 4.5

The projected ensemble indices show significant increases of 1°C and of 1.2°C in mean maximum and mean minimum temperatures respectively within the next 30 years over Baluchistan province. Summer days are projected to increase significantly by 12 days, whereas tropical nights are projected to increase significantly by 24 nights in the next 30 years over Baluchistan province. Monthly maximum value of the daily maximum temperature is also projected to increase by 0.7°C, whereas monthly minimum value of daily maximum temperature is projected to increase significantly by 1.7°C over the province. Moreover, monthly maximum of daily minimum temperature is projected to increase significantly by 0.9°C, whilst monthly minimum of daily minimum temperature is projected to increase significantly by 1.4°C over the province. Furthermore, both warm days and warm nights are projected to increase significantly by 19 days and 21 nights respectively over the next 30 years period in the Baluchistan province. It is also seen that both cold days and cold nights are projected to decrease significantly in the next 30 years projection period. Also, ensemble projections show a significant increase of 25 days in the warm spell duration index, as well as a significant decrease of 23 days in the cold spell duration index over Baluchistan province.

7.1.2.2 RCP 8.5

Both mean maximum and mean minimum temperatures, under RCP8.5 protocols, show increase of 1.1°C and 1.3°C in the next 30 years projection period over Baluchistan province (Figure 7.2). Summer days are projected to increase significantly by 26 days, whilst tropical nights are projected to increase significantly by 16 nights in the next 30 years period. Monthly maximum of daily maximum temperature is projected to increase by 1.4°C, whilst monthly maximum of daily minimum temperature is projected to increase by 1.2°C in the Baluchistan province. Both warm days and warm nights are projected to increase by 19 days and 24 nights respectively, whereas, both cool days and cool nights are projected to decrease by 13 days and 19 nights respectively in the next 30 years period. Moreover, it is seen that ensemble projections show an increase of 21 days in the warm spell duration index, and a decrease of 19 days in the cold spell duration index over the next 30 years period in the Baluchistan province.

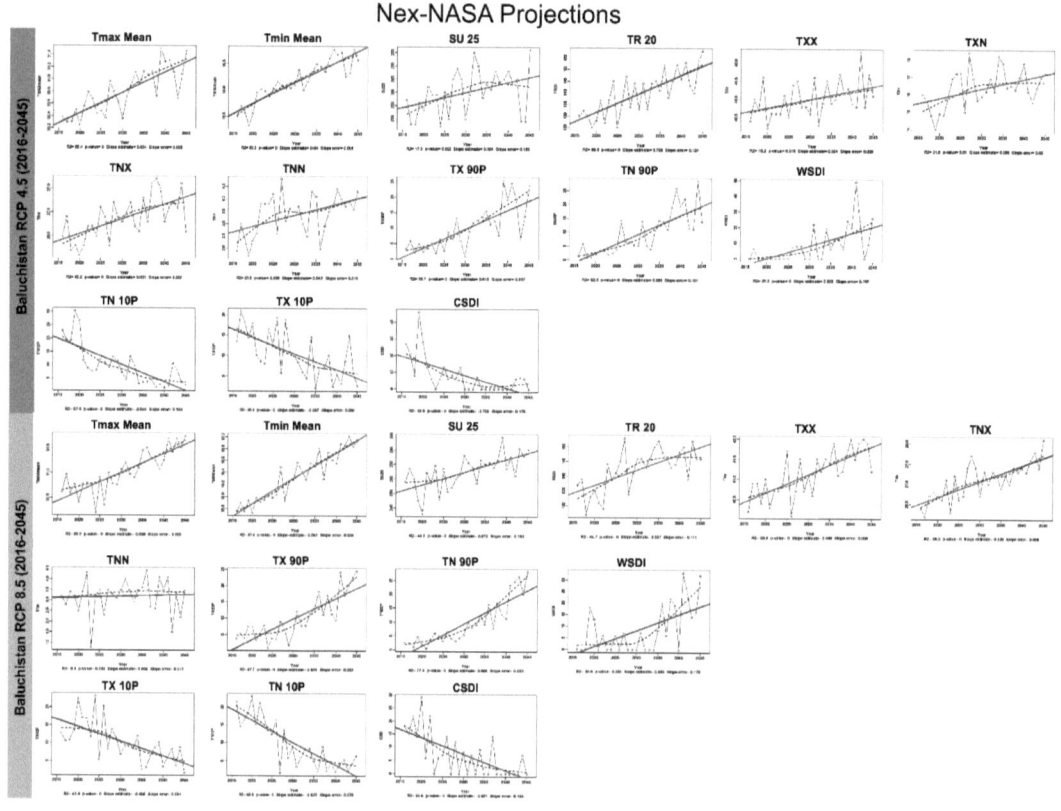

Figure 7.2: Next 30 year's (2016-2045) ensemble projected indices over Baluchistan province.

7.1.3 Projected indices over GB-AJK

7.1.3.1 RCP 4.5

Ensemble mean projected indices show increases of 1.1°C in mean maximum temperature, of 1.0°C in mean minimum temperature, of 1.1°C in monthly maximum of daily maximum temperature, of 10°C in monthly maximum of daily minimum temperature, and of 1.5°C in monthly minimum of daily minimum temperature over the GB-AJK region. Warm days are projected to increase significantly by 13 days, warm nights are projected to increase significantly by 16 nights, cool days are projected to decrease significantly by 14 days, and cool nights are projected to decrease significantly by 16 nights, in the next 30 years period over the GB-AJK region. Frost days in the GB-AJK region show a significant decrease of 18 days in the next 30 years projection period. Moreover, warm spell duration index displays 16 days increase, whilst cold spell duration index displays 20 days decrease in the next 30 years projection period over the GB-AJK region.

7.1.3.2   RCP 8.5

The high end RCP8.5 scenario projects increases of 1.5°C in mean maximum temperature, 1.6°C in mean minimum temperature, 1.2°C in monthly maximum of daily minimum temperature, and 2.3°C in monthly minimum of daily minimum temperature over the GB-AJK region (Figure 7.3). Summer days in GB-AJK region are also projected to increase significantly by 6 days in the next 30 years period. Both warm days and warm nights are projected to increase by 19 days and 24 nights respectively, whilst both cool days and cool nights are projected to decrease significantly by 17 days and 20 nights over the GB-AJK region. Mean ensemble projections show significant increase of 32 days in the warm spell duration index, while they show significant decrease of 19 days in the cold spell duration index over the GB-AJK region. The DTR is projected to decrease significantly 0.3°C in the next 30 years over the GB-AJK region. Moreover, both ice days and frost days are projected to decrease significantly by 22 days and 27 days respectively.

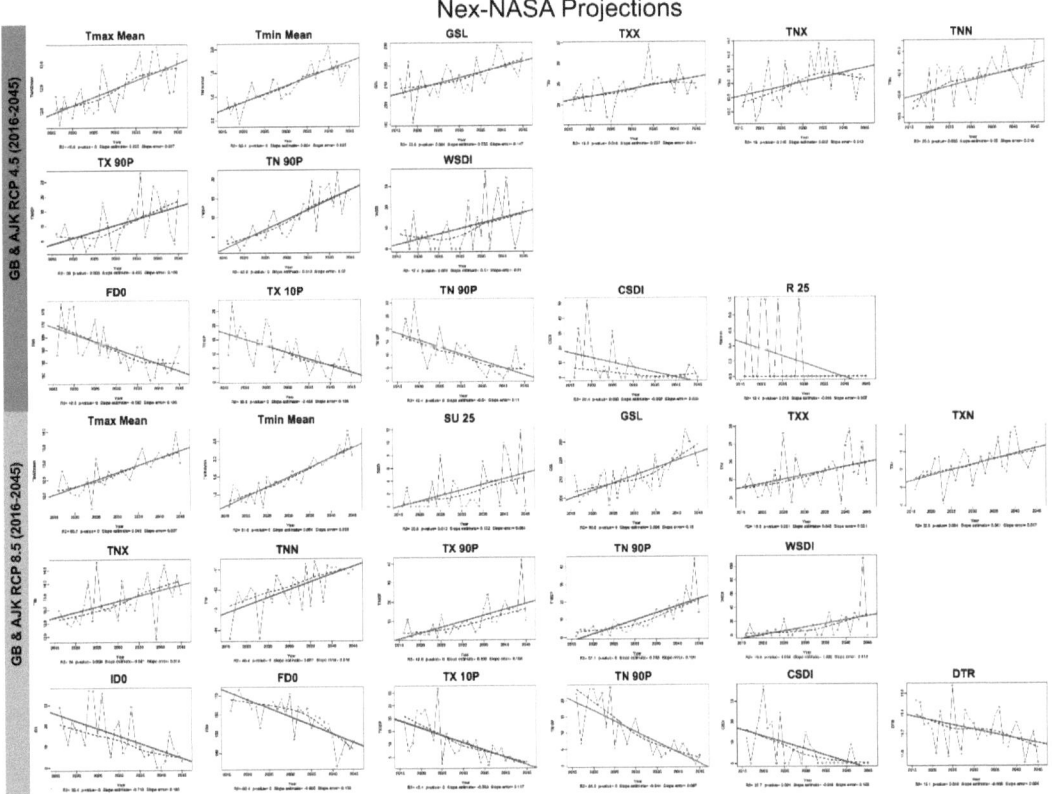

Figure 7.3: Next 30 year's (2016-2045) ensemble projected indices over GB-AJK region.

7.1.4 Projected indices over KPK

7.1.4.1 RCP 4.5

Ensemble mean projected indices show that mean maximum temperature increases by 1.1°C, mean minimum temperature increases by 1.1°C, monthly maximum of daily maximum temperature increases by 0.9°C, monthly minimum of daily maximum temperature increases by 1.6°C, monthly maximum of daily minimum temperature increases by 1.4°C, and monthly minimum of daily minimum temperature increases by 1.8°C in the next 30 years period over KPK province. Summer days are projected to increase by 18 days, whilst tropical nights are projected to increase by 29 nights over the KPK province. Both warm days and warm nights are projected to increase by 16 days and 19 nights respectively, while both cool days and cool nights are projected to decrease by 15 days and 17 nights respectively over the next 30 years in KPK province. Warm spell duration index is projected to increase by 21 days, whilst cold spell duration index is projected to decrease by 24 days in the KPK province. Moreover, frost days are projected to show a significant decrease of 17 days in the next 30 years period over the KPK province.

7.1.4.2 RCP 8.5

Mean ensemble projected indices show significant increases in mean maximum temperature by 1.3°C, in mean minimum temperature by 1.5°C, in monthly maximum of daily maximum temperature by 1.5°C, in monthly minimum of daily maximum temperature by 1.2°C, in monthly maximum of daily minimum temperature by 1.1°C, and in monthly minimum of daily minimum temperature by 2.0°C, under RCP8.5 protocols over the next 30 years period in the KPK province (Figure 7.4). Summer days are projected to increase significantly by 13 days, whilst tropical nights are projected to increase significantly by 33 days in the next 30 years projection period. Both warm days and warm nights are projected to increase significantly by days and 24 nights respectively, whereas both cool days and cool nights are projected to decrease significantly by 15 days and 20 nights respectively. Moreover, warm spell duration index projects a 26 days increase, whilst cold spell duration index projects a 28 days decrease in the next 30 years projection period over the KPK province. Furthermore, frost days are also projected to decrease significantly by 22 days in the next 30 years projection period over the KPK province.

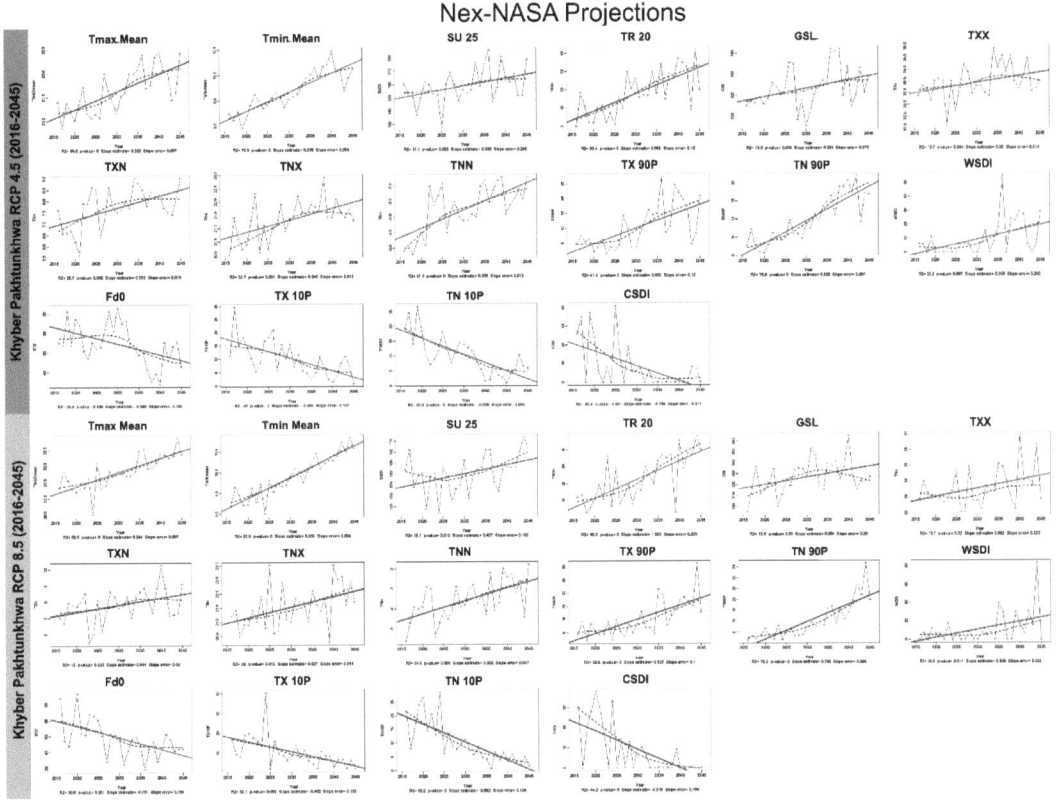

Figure 7.4: Next 30 year's (2016-2045) ensemble projected indices over KPK province.

7.1.5 Projected indices over Sindh

7.1.5.1 RCP 4.5

Projected indices from mean ensemble Nex-NASA product show increases of 0.8°C in mean maximum temperature, 1.0°C in mean minimum temperature, 1.3°C in monthly maximum of daily maximum temperature, 1.3°C in monthly minimum of daily maximum temperature, and 1.0°C in monthly minimum of daily minimum temperature over the next 30 years period in the Sindh province. Summer days are projected to increase significantly by 16 days, whereas tropical nights are projected to increase significantly by 10 nights in the next 30 years projection period. Both warm days and warm nights project significant increases of 17 days and 20 nights respectively, whilst both cold days and cold nights project significant decreases of 11 days and 21 nights respectively. Warm spell duration index projects 22 days increase, while cold spell duration index displays 21 days decrease over the Sindh province. Interestingly, there is a projected increase of 19 mm in the 5 days consecutive precipitation amount in the Sindh province. Moreover, the consecutive wet days in Sindh also project a significant increase of 5 days in the next 30 years projection period.

7.1.5.2 RCP 8.5

Mean maximum temperature and mean minimum temperature show projected increases of 1.0°C and 1.2°C under RCP8.5 protocols over the Sindh province (Figure 7.5). Summer days are projected to increase significantly by 14 days, whilst tropical nights are projected to increase significantly by 19 days in the next 30 years projection period. Monthly maximum of daily minimum temperature is projected to increase by 0.9°C in the Sindh province. Moreover, both warm days and warm nights project significant increase of 19 days and 25 nights respectively, whereas both cold days and cold nights project significant decreases of 13 days and 19 nights respectively, in the Sindh province. Furthermore, the warm spell duration index projects significant increase of 20 days, whilst the cold spell duration index displays a significant decrease of 14 days, in the next 30 years projection period over the Sindh province.

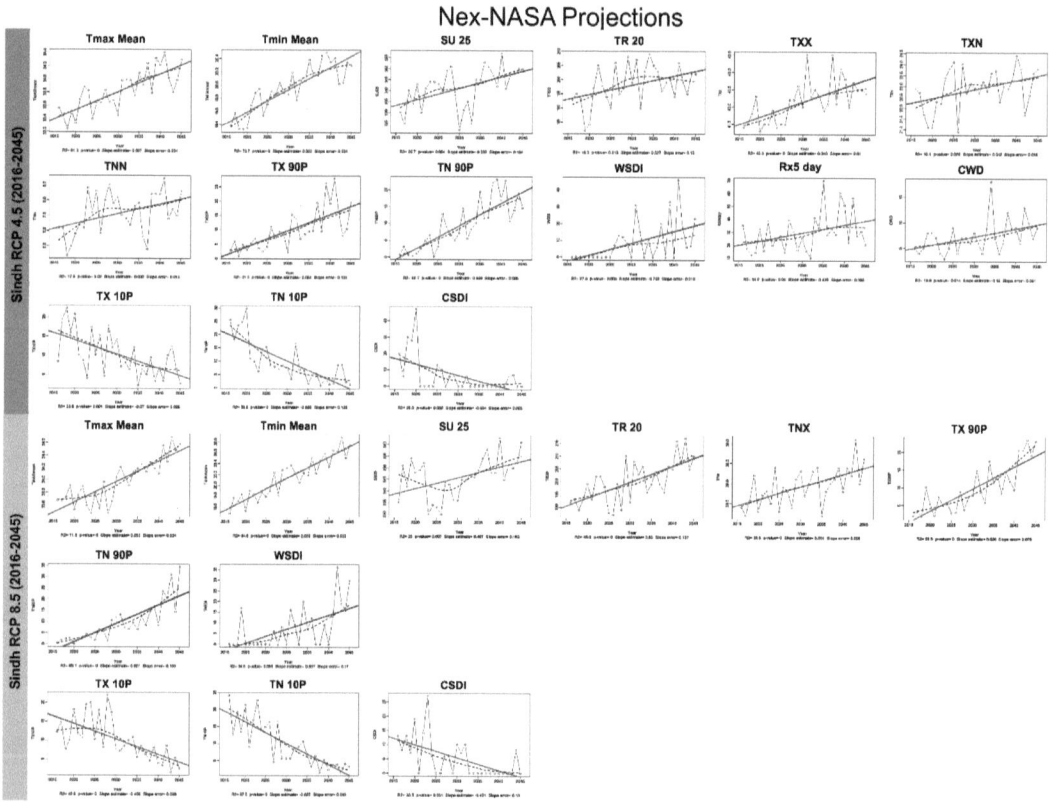

Figure 7.5: Next 30 year's (2016-2045) ensemble projected indices over Sindh province.

7.2 PROJECTED CLIMATE OVER PAKISTAN BY 2050

7.2.1 Projected climate in GB-AJK

Normal precipitation in GB-AJK region for DJF is 1.4 mm/day as simulated by Nex-NASA over 1970-2000 historical period (Table 7.1). Under RCP4.5, it is projected to increase by 17%, and under RCP8.5, it is projected to increase by 16%, over the 2010-2050 projection period. The normal simulated precipitation in MAM is shown to be 2.0 mm/day which is projected to increase by 11% in 2010-2050 projection period. For JJA, the normal simulated precipitation is seen to be 2.0 mm/day which is projected to increase by 14% under RCP4.5 and by 13% under RCP8.5 in the 2010-2050 projection period. The simulated normal precipitation in SON is seen to be 0.8 mm/day, and is projected to increase by 12% under RCP4.5 and by 9% under RCP8.5 over the 2010-2050 projection period.

The normal simulated maximum temperature in the GB-AJK for DJF is seen to be 0.5°C, which is seen to increase by 1.4°C under RCP4.5 and by 1.7°C under RCP8.5 over the 2010-2050 projection period. In MAM, the normal simulated maximum temperature is 10.4°C, which is projected to increase by 1.4°C under RCP4.5 and by 1.6°C under RCP8.5. Normal simulated maximum temperature in JJA is 20.6°C which is projected to increase by 1.1°C under RCP4.5 and by 1.5°C under RCP8.5 over the projection period. The SON normal simulated maximum temperature is seen to be 12.9°C which is seen to increase by 1.4°C under RCP4.5 and by 1.7°C under RCP8.5 over the 2010-2050 projection period in the GB-AJK region.

Simulated normal minimum temperature (1970-2000) in DJF is -9.7°C in the GB-AJK region, which is projected to increase by 1.5°C and 1.8°C under RCP4.5 and RCP8.5 respectively. In MAM, simulated mean minimum temperature is -0.7°C which is projected to increase by 1.5°C and 1.8°C under RCP4.5 and RCP8.5 respectively. The JJA simulated normal minimum temperature is 8.8°C which is projected by increases of 1.5°C and 1.8°C under RCP4.5 and RCP8.5 respectively. In the SON, simulated normal minimum temperature is 0.2°C, which is projected to increase by 1.7°C under RCP4.5 and by 1.9°C under RCP8.5 over the GB-AJK region.

7.2.2 Projected climate in KPK

Simulated normal precipitation in KPK for DJF is 1.5 mm/day, which projects an increase of 16% under RCP4.5 and an increase of 17% under RCP8.5 in the 2010-2050 projection period. For MAM, the normal simulated precipitation is 2.2 mm/day which is shown to project increases of 5.4 mm/day under RCP4.5 and of 5.6 mm/day under RCP8.5. The JJA simulated normal precipitation is 2.1 mm/day, which is suggested to project increases of 16% and 17% under RCP4.5 and RCP8.5 respectively. Simulated normal precipitation in SON is 0.8 mm/day which project increases of 16% and 13% under RCP4.5 and RCP8.5 respectively.

Simulated normal maximum temperature in DJF over KPK is 9.8°C which project increases of 1.4°C and 1.7°C under RCP4.5 and RCP8.5 respectively. In MAM, the simulated normal maximum temperature is 20.1°C, with projected increases of 1.6°C under RCP4.5 and 1.9°C under RCP8.5. Simulated normal maximum temperature in JJA is 29.3°C, with projected increases of 0.9°C under RCP4.5 and 1.3°C under RCP8.5 in the KPK province. The SON has the simulated normal maximum temperature value of 21.7°C, with projected increases of 1.4°C and 1.6°C under RCP4.5 and RCP8.5 respectively.

Simulated normal minimum temperature in DJF over the KPK province is -2.0°C with projected increases of 1.6°C and 1.9°C under RCP4.5 and RCP8.5 respectively. The MAM displays a simulated normal minimum temperature value of 8.2°C with projected increases of 1.6°C and 1.9°C under RCP4.5 and RCP8.5 respectively. The JJA displays a normal simulated minimum temperature of 17.6°C, which project increases of 1.2°C under RCP4.5 and of 1.5°C under RCP8.5 in the KPK province. In SON, the simulated normal minimum temperature is 8.6°C with projected increases of 1.6°C and 1.9°C under RCP4.5 and RCP8.5 respectively.

7.2.3 Projected climate in Punjab

The simulated normal precipitation over Punjab in DJF is 0.5 mm/day, in MAM is 0.9 mm/day, in JJA is 2.3 mm/day, and in SON is 0.5 mm/day, with projections of 12% increase in DJF, of 1% increase in MAM, of 11% increase in JJA, and of 22% increase in SON, under RCP4.5, whereas under RCP8.5, it projects an increase of 11% in the DJF, decrease of 2% in the MAM, increase of 13% in JJA, and increase of 15% in SON over the 2010-2050 projection period. The simulated normal maximum temperature in DJF is 19.9°C, in MAM is 31.4°C, in JJA is 36.9°C, and in SON is 30.4°C, with projected increases of 1.4°C (1.6°C) in the DJF, of 1.5°C (1.8°C) in the MAM, of 0.8°C (1.0°C) in the JJA, and of 1.2°C (1.4°C), under RCP4.5 (RCP8.5) in the SON over the Punjab province. It is also seen that simulated normal minimum temperature over the province is 5.2°C in the DJF, 16.9°C in the MAM, 25.5°C in the JJA, and 15.8°C in the SON, with projected increases of 1.4°C (1.7°C) in the DJF, 1.5°C (1.8°C) in the MAM, 1.0°C (1.3°C) in the JJA, and 1.5°C (1.8°C) in SON, under RCP4.5 (RCP8.5), over the Punjab province.

7.2.4 Projected climate over Sindh

Simulated normal precipitation over Sindh in DJF is 0.1 mm/day, in MAM is 0.0 mm/day, in JJA is 1 mm/day, and in SON is 0.1 mm/day, with projections of 12% increase in the DJF (RCP4.5), of 14% increase in the MAM (RCP4.5), of 12% increase in the JJA (RCP4.5), and of 17% increase in the SON (RCP4.5), while under RCP8.5, it projects decrease of 2.5% in the DJF, increase of 21% in the MAM, increase of 10% in the JJA, and substantial increase of 58% in the SON over the 2010-2050 projection period. The normal simulated maximum temperature in Sindh is 25.0°C in DJF, is 35.4°C in MAM, is 37.0°C in JJA, and is 33.1°C in SON, with projected increases of 1.4°C (1.7°C) in the DJF, of 1.2°C (1.5°C) in the MAM, of 0.7°C (1.0°C) in the JJA, and of 1.3°C (1.3°C) in the SON, under RCP4.5 (RCP8.5), over the 2010-2050 projection period. Moreover, it is seen that simulated normal minimum temperature over the Sindh province is 9.1°C in the DJF, 20.0°C in the MAM, 26.4°C in the JJA, and 18.6°C in the SON, with projected increases of 1.6°C (1.9°C) in the DJF, 1.3°C (1.6°C) in the MAM, 0.9°C (1.2°C) in the JJA, and 1.6°C (1.8°C) in the SON, under RCP4.5 (RCP8.5) projection period.

7.2.5 Projected climate in Baluchistan

Simulated normal precipitation over Baluchistan is 0.4 mm/day in DJF, is 0.3 mm/day in MAM and in JJA, and is 0.0 mm/day in SON, with projected increases of 10% (3%) in the DJF, 6% (5%) in the MAM, 12% (11%) in the JJA, and 16% (30%) in the SON, under RCP4.5 (RCP8.5) over the 2010-2050 projection period. In terms of maximum temperature, the simulated normal is 19.2°C for DJF, is 30.5°C for MAM, is 37.7°C for JJA, and is 30.1°C for SON, with projected increases of 1.5°C (1.8°C) in the DJF, 1.3°C (1.6°C) in the MAM, 1.0°C (1.3°C) in the JJA, and 1.6°C (1.7°C) in the SON, under RCP4.5 (RCP8.5) projection period. Furthermore, it is seen that simulated normal minimum temperature is 4.9°C in the DJF, is 15.8°C in the MAM, is 23.9°C in the JJA, and is 13.8°C in the SON, with projections of increase by 1.6°C (1.9°C) in the DJF, by 1.3°C (1.6°C) in the MAM, by 1.2°C (1.5°C) in the JJA, and by 1.8°C (2.0°C) in the SON, under RCP4.5 (RCP8.5) over the projection period.

7.2.6 Projected climate over All Pakistan

Whole Pakistan is seen to display a normal simulated precipitation value of 0.6 mm/day in the DJF, of 0.7 mm/day in the MAM, of 2.0 mm/day in the JJA, and of 0.6 mm/day in the SON, with projected increases of 13% (13%) in the DJF, 10% (9%) in the MAM, 3% (6%) in the JJA, and 21% (21%) in the SON, under RCP4.5 (RCP8.5). Simulated normal maximum temperature is 15°C in the DJF, is 25.5°C in the MAM, is 31.1°C in the JJA, and is 24.7°C in the SON, with projected increase of 1.4°C (1.6°C) in the DJF, of 1.3°C (1.6°C) in the MAM, of 1.0°C (1.2°C) in the JJA, and 1.3°C (1.5°C) in the SON, under RCP4.5 (RCP8.5) projection period. Moreover, it is seen that for whole Pakistan, simulated normal minimum temperature is 2.9°C for DJF, is 12.9°C for MAM, is 19.9°C for JJA, and is 11.7°C for SON, with projected increases of 1.6°C (1.8°C) in the DJF, 1.4°C (1.6°C) in the MAM, 1.2°C (1.4°C) in the JJA, and 1.5°C (1.7°C) in the SON, under RCP4.5 (RCP8.5) projection period. Through the analysis, it is seen that SON precipitation has displayed significant increase by 21% over the whole Pakistan. Projections over the upper Sindh (which has a history of prolonged droughts), are displaying significant increase of 57% in precipitation regime. Also, we have found that projected seasonal minimum temperatures changes are overall higher than the projected seasonal maximum temperature changes over the whole Pakistan.

Table 7.1: Historical and projected seasonal climate over Pakistan

Region	Scenario	Precipitation (mm)				TMAX (°C)				TMIN (°C)			
		DJF	MAM	JJA	SON	DJF	MAM	JJA	SON	DJF	MAM	JJA	SON
GB-AJK	Historical (1970-2000)	1.4	2.0	2.0	0.8	0.5	10.4	20.6	12.9	-9.7	-0.7	8.8	0.2
	RCP4.5 (2010-2050)	1.6	2.2	2.3	0.9	1.9	11.8	21.7	14.3	-8.2	0.8	10.2	1.8
	RCP8.5 (2010-2050)	1.6	2.2	2.3	0.9	2.2	12.0	22.1	14.6	-7.9	1.1	10.6	2.1
KPK	Historical (1970-2000)	1.5	2.2	2.1	0.8	9.8	20.1	29.3	21.7	-2.0	8.2	17.6	8.6
	RCP4.5 (2010-2050)	1.7	2.4	2.4	1.0	11.2	21.7	30.2	23.1	-0.4	9.8	18.8	10.2
	RCP8.5 (2010-2050)	1.7	2.4	2.4	0.9	11.5	22.0	30.5	23.3	-0.1	10.0	19.2	10.4
Punjab	Historical (1970-2000)	0.5	0.9	2.3	0.5	19.9	31.4	36.9	30.4	5.2	16.9	25.5	15.8
	RCP4.5 (2010-2050)	0.6	0.9	2.5	0.7	21.2	33.0	37.7	31.7	6.6	18.4	26.5	17.4
	RCP8.5 (2010-2050)	0.6	0.9	2.5	0.6	21.5	33.3	38.0	31.8	6.8	18.7	26.8	17.6
Sindh	Historical (1970-2000)	0.1	0.0	1.0	0.1	25.0	35.4	37.0	33.1	9.1	20.0	26.4	18.6
	RCP4.5 (2010-2050)	0.1	0.0	1.1	0.1	26.4	36.6	37.7	34.4	10.7	21.3	27.3	20.3
	RCP8.5 (2010-2050)	0.1	0.0	1.1	0.2	26.7	36.8	38.0	34.4	11.0	21.6	27.5	20.4
Baluchistan	Historical (1970-2000)	0.4	0.3	0.3	0.0	19.2	30.5	37.7	30.1	4.9	15.8	23.9	13.8
	RCP4.5 (2010-2050)	0.5	0.3	0.4	0.0	20.7	31.8	38.7	31.6	6.5	17.1	25.1	15.6
	RCP8.5 (2010-2050)	0.5	0.3	0.3	0.0	21.1	32.1	39.1	31.8	6.7	17.4	25.3	15.8
All Pakistan	Historical (1970-2000)	0.6	0.7	2.0	0.6	15.0	25.5	31.1	24.7	2.9	12.9	19.9	11.7
	RCP4.5 (2010-2050)	0.7	0.8	2.1	0.7	16.4	26.8	32.1	26.0	4.5	14.2	21.1	13.2
	RCP8.5 (2010-2050)	0.7	0.8	2.1	0.7	16.7	27.1	32.4	26.2	4.7	14.5	21.4	13.4

8 Discussion

As per observed data analysis, summer days in the GB-AJK are increasing significantly which may trigger high frequencies of Glacier Lake Outburst Floods in those regions. Significant increasing trends in very heavy precipitation extremes are seen in the KPK province which may bring flood related liabilities in the region. This work has seen high confidence in the trend increase of minimum temperature related indices in the Punjab province, which implicates agricultural hazards to the food security of the country. The mean maximum temperatures, the summer days, and the warm days, all display significant increasing trends in the Sindh province, which may evolve high frequency and prolonged heat waves in the region. Last but not least, significant increasing trends in temperature related indices are liable to bring thermal instability in the Baluchistan province, which ultimately brings the potential of high frequency dust storms over the region (detailed trends for all analyzed climate indices may be seen in Appendix 1).

Pakistan has been facing shortages in both power and water sector which are lifelines of a country. Significant increases in the maximum and minimum temperatures over the country may affect such sectors drastically. This report shall bring added value to all the stake holders and the policy maker in determining the hazards that extreme climate has brought in the past and may bring in the near future.

9 Conclusions

Based on the observed and model based analysis of the core extreme climate indices, following conclusions are drawn:

9.1 GB-AJK

Both observed and models based analyses have shown significant increasing trends in the mean maximum temperatures of the GB-AJK region in the past. Increase in mean maximum temperature over Gilgit is 1.5°C and over Muzaffarabad is 1.1°C in the past 54 years. Projected climate has suggested an added increase of 1.4°C in the mean maximum temperature over the GB-AJK region in the next 30 years (by year 2045). Summer days have shown significant increases in Gilgit and Muzaffarabad by 24 days and 18 days respectively. Projected climate has suggested an added increase of 6 days in the summer days index over the GB-AJK region in the next 30 years (by the year 2045). This ultimately suggests that winters are becoming shorter in the GB-AJK region and may continue to follow suit in the near future.

9.2 KPK

High confidence is attained through observed and model based constructions of mean maximum temperature index over the KPK region. Peshawar has shown a 0.9°C increase in the past index, whereas, the KPK province has further been projected to show a whopping 1.3°C added increase in the mean maximum temperature by the year 2045. Warm days in Dir have shown significant increase of 16 days in the past index, whereas, the KPK province has further been projected to show an added 17 days increase in the index by the year 2045.

9.3 Punjab

Mean minimum temperatures in the Punjab province have shown significant increases in the past trends. Average increase in the mean minimum temperatures is 1.9°C over the past 54 years over the Punjab province. Projected index for the mean minimum temperature has suggested an added 1.4°C increase in the next 30 years (by the year 2045) over the Punjab province. Warm nights index in the Punjab province has also shown significant increase of 12 nights over the past 54 years historical period. Projected value for the index suggests an added increase of 25 nights over the next 30 years in the Punjab province. Increase in minimum temperature indices over the Punjab province may bring liabilities for food security in the region.

9.4 Sindh

The Sindh province has seen a 1.6°C increase in both the mean maximum and the mean minimum temperature indices over the past 54 years historical period. Projected indices for mean maximum temperature suggests an added 1.0°C increase, whilst for mean minimum temperature suggests an added 1.2°C increase over the next 30 years projection period. No. of heat waves days in Karachi have increased by 8 days over the historical period, while they are suggested to increase further by 20 days in the next 30 years projection period over the Sindh province. Heat waves have claimed lives and have brought extreme distress among the masses over the Sindh province. Tremendous projected increase in the heat waves over the province seeks special attention towards future policy making.

9.5 BALUCHISTAN

Mean maximum temperature index in the Baluchistan province has shown an increase of 1.8°C, whilst mean minimum temperature index in the province has shown an increase of 1.9°C over the past 54 years historical period. Projections suggest an added 1.1°C increase in the mean maximum temperature, and an added 1.3°C increase in the mean minimum temperatures over the next 30 years period in the province. Both the observed and the model based calculated indices have shown significant increases of 29 days in the tropical nights, 11 days in the warm days, and 10 nights in the warm nights, over the past 54 years historical period. Projected analysis suggests added increases of 16 days in the tropical nights, 19 days in the warm days, and 24 nights in the warm night indices over the next 30 years projected period in the Baluchistan province.

10 REFERENCES

Burhan A , Waheed I , Syed AABRasul G, Shreshtha AB and Shea JM., 2015. Generation of High–Resolution Gridded Climate Fields for the Upper Indus River Basin by Downscaling CMIP5 Outputs. J Earth Sci Clim Change 6:2

Burhan, A., G. Rasul , W. Iqbal , S. A. A. Bukhari 2014. Regional Comparison between Global Circulation Model GCM20 and Regional Climate Model PRECIS. Pakistan Journal of Meteorology Vol. 11, Issue 21: Jul, 2014 39

Burhan, A., S. Haider , S. A. A. Bukhari 2015. Regional Precipitation Response to Regional Warming in Past and Future Climate. Pakistan Journal of Meteorology Vol. 11, Issue 22: Jan, 2015 57.

Chaudhry, Qamar-ul-Zaman, et al. 2009. Climate Change Indicators of Pakistan. Islamabad: Pakistan Meteorological Department, 2009. PMD-22/2009.

Choi, G. et al., 2009. Changes in means and extreme events of temperature and precipitation in the Asia-Pasific Network region, 1955-2007. *International Journal of Climatology*, pp. 1906-1925.

Frich, P. et al., 2002. Observed coherent changes in climatic extremes. *Climate Research*, pp. 193-212.

Houghton, J. T. et al., 2001. *Climate Change 2001: The Scientific Basis*, New York: Intergovernmental Panel on Climate Change.

Karl, T.R., N. Nicholls, and A. Ghazi, 1999. CLIVAR/GCOS/WMO workshop on indices and indicators for climate extremes: Workshop summary. Climatic Change, 42, 3-7.

Klein Tank, A. M. G. & Konnen, G. P., 2003. Trends in Indices of Daily Temperature and Precipitation Extremes in Europe, 1946-99. *Journal of Climate*, pp. 3665-3680.

Lynn, B. H., Healy, R. & Druyan, L. M., 2007. An Analysis of the Potential for Extreme Temperature Change Based on Observations and Model Simulations. *Journal of Climate*, Volume 20, pp. 1539-1554.

McSweeney, C. F., Jones, R. G., Lee, R. W. & Rowell, D. P., 2015. Selecting CMIP5 GCMs for downscaling over multiple regions. *Clim Dyn,* pp. 3237-3260.

Meehl, G. A. & Claudia, T., 2004. More Intense, More Frequent, and Longer Lasting Heat Waves in the 21st Century. *Science,* 13 August, pp. 994-997.

Meehl, G. A. et al., 2000. Trends in Extreme Weatherr and Climate Events: Issues RElated to Modeling Extremes in Projections of Future Climate Change. *Bulletin of the American meteorological Society,* 81(3), pp. 427-436.

Peterson, T. C. et al., 2001. *Report on the Activities of the Working Group on Climate Change Detection and Related Rapporteure,* Geneva: World Meteorological Organization.

Salma, S., S. Rehman, M. A. Shah., 2012. Rainfall Trends in Different Climate Zones of Pakistan. Pakistan Journal of Meteorology Vol. 9, Issue 17

Sillmann, J. & Roeckner, E., 2008. Indices for extreme events in projections of anthropogenic climate change. *Climatic Change,* pp. 83-104.

Sillmann, J. et al., 2013. Climate extremes indices in the CMIP5 multimodel ensemble: Part 1. Model evaluation in the present climate. *Journal of Geophysical Research: Atmospheres,* pp. 1716-1733.

Sillmann, J. et al., 2013. Climate extremes indices in the CMIP5 multimodel ensemble: Part 2 Future climate projections. *Journal of Geophysical Research: Atmospheres,* pp. 2473-2493.

Taylor, K. E., 2001. Summarizing multiple aspects of model performance in a single diagram, J. Geophys. Res., 106(D7), 7183–7192 - See more at: https://climatedataguide.ucar.edu/climate-data-tools-and-analysis/taylor-diagrams#sthash.hAaC5Lba.dpuf

Taylor, K. E., R. J. Stouffer, and G. A. Meehl, 2012. An overview of CMIP5 and the experiment design. Bull. Am. Meteorol. Soc., 93, 485–498.

Tebaldi, C., Hayhoe, K., Arblaster, J. M. & Meehl, G. A., 2006. Going to the Extremes - An Intercomparision of Model-Simulated Historical and Future Changes in Extreme Events. *Climatic Change*, pp. 185-211.

Thrasher, B., E. P. Maurer, C. McKellar, and P. B. Duffy., 2012. Technical Note: Bias correcting climate model simulated daily temperature extremes with quantile mapping Hydrol. Earth Syst. Sci., 16, 3309–3314, www.hydrol-earth-syst-sci.net/16/3309/2012/ doi:10.5194/hess-16-3309-2012 © Author(s) 2012. CC Attribution 3.0 License.

William R. Goomn, L gregory J. Mcrae, and John H. Seinfeld., 1979. A Comparison of Interpolation Methods for Sparse Data: Application to Wind and Concentration Fields. American Meteorological Society. 00218952/79/06076111$06.75.

Zhang, X., et al., 2005. Avoiding Inhomogeneity in Percentile-Based Indices of Temperature Extremes. J. Climate, 18, 1641-1651.

11 APPENDIX 1

11.1 Observed climate indices (1960-2013)

Stations\Index	TMAXmean	TMINmean	su25	id0	tr20	fd0	txx	txn	tnx	tnn	tx10p	tx90p	tn10p	tn90p	wsdi	csdi	dtr	rx1day	rx5day	sdii	r10mm	r20mm	cdd	cwd	r95p	r99p	prcptot
Gilgit-Baltistan and Azad Jammu Kashmir																											
Gilgit	0.028	-0.013	0.446	0	-0.288	-0.025	0.002	0.032	-0.076	0.008	-0.119	0.194	0.05	-0.118	0.135	0.044	0.041	-0.193	-0.334	-0.075	-0.005	-0.023	0.064	0.013	-1.946	-1.504	-1.308
Gupis	0.02	-0.038	0.265	-0.06	-0.659	0.379	-0.017	0.048	-0.086	-0.018	-0.085	0.133	0.133	-0.302	0.189	0.099	0.058	0.15	0.519	0.1	0.109	0.043	0.463	0.005	0.819	0.526	2.88
Kotli	0.003	-0.019	0.243	0	-0.098	0.048	-0.005	-0.011	-0.059	-0.022	0.004	0.032	0.101	-0.16	0.072	0.113	0.023	0.883	0.892	0.105	-0.008	0.106	0.062	-0.112	4.168	2.842	1.223
Muzaffarabad	0.02	-0.001	0.321	0	-0.019	0.047	0.012	0.044	-0.025	-0.019	-0.011	0.16	0.01	-0.027	0.097	0.044	0.02	0.992	1.773	0.124	0.397	0.326	-0.071	-0.058	7.979	3.68	11.734
Skardu	0.043	-0.018	0.488	-0.235	-0.145	0.179	0.016	0.084	-0.057	0.014	-0.214	0.254	0.096	-0.068	0.192	-0.021	0.058	0.145	0.195	0.005	0.062	0.012	-0.599	0.019	0.352	0.067	1.763
Khyber Pakhtun																											
Cherat	0.003	-0.024	0.2	0.001	-0.782	-0.047	0.01	-0.028	-0.061	0.027							0.047	-0.004	0.38	0.037	0.015	-0.04	-0.063	-0.001	-0.802	0.275	-0.964
DI Khan	0	0.001	0.123	0	0.254	-0.003	0.013	-0.06	-0.027	0.009	-0.023	-0.025	-0.093	-0.082	-0.048	-0.264	-0.003	0.164	0.259	0.035	0.063	0.058	-0.422	-0.003	1.681	0.29	2.732
Dir	0.028	-0.01	0.323	0	-0.349	0.036	0.01	0	-0.036	0.009	-0.059	0.285	0.009	-0.085	0.285	-0.075	0.041	0.551	1.51	-0.003	0.032	0.068	0.152	0.017	1.778	1.185	1.839
Parachinar	-0.003	-0.062	-0.066	0.002	-0.249	1.144	0.001	0	-0.034	-0.099	0.166	0.091	0.369	-0.084	0.21	0.6	0.057	-0.383	-0.395	-0.076	0.009	-0.013	0.085	0.025	-2.239	-0.963	0.793
Peshawar	0.016	0.017	0.423	0	0.336	-0.005	0.006	-0.008	-0.023	0.021	-0.064	0.053	-0.107	0.135	0.051	-0.13	-0.004	0.672	1.133	0.067	0.161	0.111	-0.15	0.013	2.82	1.053	4.846
Punjab																											
Bahawalnagar	0.014	0.042	0.263	0	0.648	-0.046	0.036	-0.094	0.002	0.05	-0.033	0.162	-0.347	0.297	0.206	-0.618	-0.014	0.318	0.522	-0.019	0.104	0.053	-1.502	0.02	0.513	0.351	2.999
Islamabad	0.018	0.02	0.36	0	0.402	-0.13	0.005	-0.002	-0.018	0.018	-0.09	0.108	-0.198	0.087	0.09	-0.314	-0.004	0.796	1.324	0.03	0.07	0.07	0.024	-0.004	3.092	2.074	4.62
Khanpur	0	0.029	0.147	0	0.425	-0.06	0.023	-0.016	0.014	0.038	0.014	-0.049	-0.195	0.192	-0.126	-0.233	-0.031	0.648	1.28	-0.011	0.018	0.014	0.08	0.024	1.297	0.771	1.556
Lahore	-0.009	0.049	0.186	0	0.651	-0.027	-0.009	-0.038	0.017	0.081	-0.016	-0.111	-0.425	0.322	-0.125	-0.771	-0.058	0.096	0.562	0.034	0.112	0.049	-0.302	0.004	1.518	0.951	2.843
Multan	-0.001	0.037	0.241	0	0.382	-0.045	0.025	-0.071	-0.019	0.058	0.005	-0.016	-0.239	0.235	-0.119	-0.354	-0.038	-0.455	-0.31	-0.041	0.015	-0.002	-0.178	0.008	-0.413	-0.895	-0.095
Sialkot	-0.004	0.034	0.277	0	0.725	0.001	0	-0.058	-0.006	-0.002	-0.024	-0.105	-0.26	0.248	-0.187	-0.271	-0.027	-0.941	-0.806	-0.041	0.069	0.062	-0.063	0.003	-2.146	-0.911	0.437
Sindh																											
Badin	-0.011	0.031	-0.121	0	0.452	0	-0.008	-0.004	0.006	0.053	0.009	-0.236	-0.294	0.184	-0.661	-0.507	-0.039	-0.165	0.214	-0.087	0.006	0.021	-0.842	0.041	0.324	0.368	0.923
Chor	0.028	0.017	0.209	0	0.259	-0.039	0.041	0.012	0.006	0.05	-0.214	0.16	-0.12	0.092	-0.02	-0.168	0.014	0.813	1.665	0.077	0.024	0.032	-0.084	0.02	1.619	1.175	2.299
Karachi	0.027	0.038	0.48	0	0.672	0	0.027	0.048	0.015	0.048	-0.307	0.138	-0.293	0.29	0.107	-0.835	0.041	0.091	-0.243	-0.015	-0.036	-0.014	0.392	0.004	-0.771	-0.004	-0.838
Nwabshah	0.033	0.002	0.366	0	0.127	0.008	0.065	0.043	-0.002	-0.003	-0.238	0.2	-0.011	-0.06	0.144	0.006	0.032	0.051	0.233	0.052	0.019	0.011	-0.476	0.004	0.23	-0.146	0.445
Rohri	0.008	0.019	0.292	0	0.301	-0.001	0.023	-0.003	-0.01	0.003	-0.05	0.065	-0.097	0.257	0.098	-0.212	-0.014	0.509	0.551	0.112	0.011	0.018	0.266	-0.013	0.717	0.673	0.185
Baluchistan																											
Barkhan	0.025	-0.037	0.588	0	-0.56	0.068	0.012	0.03	-0.031	-0.032	-0.117	0.183	0.361	-0.112	0.296	0.817	0.063	-0.001	-0.291	0.031	0.06	0.056	0.6	-0.027	-0.08	-0.786	0.111
Dalbandin	0.05	0.029	0.592	0	0.517	-0.322	0.045	0.022	0	-0.22	-0.23	0.291	-0.189	0.074	0.322	-0.303	0.016	-0.475	-0.653	-0.046	0.004	-0.001	1.033	0.013	-0.872	-0.966	-0.886
Khuzdar	0.034	0.028	0.665	0	0.605	-0.028	0.043	0.043	-0.021	0.061	-0.171	0.173	-0.106	0.267	0.229	0.023	0	0.175	0.45	0.069	0.118	0.065	0.308	-0.007	0.754	0.094	2.119
Panjgur	0.026	0.025	0.416	0	0.574	-0.088	0.009	0.006	-0.014	0.02	-0.124	0.165	-0.204	0.096	0.171	-0.271	-0.001	-0.291	-0.285	-0.057	-0.005	-0.002	-0.161	-0.011	-0.314	-0.302	-0.466
Quetta	0.038	0.052	0.528	-0.009	0.491	-0.508	0.028	0.018	0.015	0.042	-0.2	0.208	-0.421	0.163	0.148	-0.661	-0.014	0.242	0.032	0.002	0.003	0.02	0.261	0.037	0.374	0.252	0.693
Zhob	0.025	-0.013	0.471	0	-0.366	0.001	0.01	0.019	-0.053	-0.018	-0.089	0.148	0.137	-0.062	0.141	0.258	0.041	-0.278	-0.465	-0.013	0.018	0.007	-0.024	-0.004	-0.07	-0.663	0.136
Pakistan	0.015	0.017	0.265	0	0.323	-0.08	0.003	0.027	-0.014	0.034	-0.13	0.059	-0.252	0.077	0.113	-0.537	-0.002	0.109	0.163	0	0.038	0.01	0.111	0.472	0.717	0.587	0.111

Bold P<0.05
Slope>0
Slope<0

11.2 Ensemble mean GCMs simulated climate indices (1970-2004)

Stations\Index	TMAXmean	TMINmean	su25	id0	tr20	fd0	txx	txn	tnx	tnn	tx10p	tx90p	tn10p	tn90p	wsdi	csdi	dtr	rx1day	rx5day	sdii	r10mm	r20mm	cdd	cwd	r95p	r99p	prcptot
GB & AJK	0.026	0.024	0.174	-0.336	0	-0.098	0.032	0.074	0.024	0.082	-0.195	0.237	-0.203	0.174	0.207	-0.152	0.002	0.046	-0.622	0.003	0.022	-0.041	0.108	0.084	-0.785	0.118	0.836
Khyber Pakhtunkhwa	0.021	0.028	0.001	-0.015	0	-0.174	-0.007	0.079	0.022	0.094	-0.202	0.073	-0.264	0.252	0.039	-0.345	-0.007	0.057	-0.458	0.006	0.018	-0.015	0.126	-0.034	0.047	-0.444	0.629
Punjab	0.023	0.022	0.303	0	0.133	-0.134	0.009	0.065	0.016	0.058	-0.167	0.17	-0.239	0.25	0.291	-0.288	0.001	-0.169	0.058	-0.001	0.046	-0.031	-0.22	0.301	-0.541	-0.851	2.167
Sindh	0.021	0.022	0.513	0	0.171	0	-0.013	0.061	0.025	0.05	-0.226	0.171	-0.309	0.29	0.29	-0.42	-0.001	-0.072	-0.003	0.026	0.152	0.033	1.18	0.266	1.384	-0.062	4.61
Baluchistan	0.02	0.022	0.211	0	0.321	-0.174	0.028	0.067	0.02	0.026	-0.176	0.17	-0.251	0.288	0.274	-0.233	-0.002	-0.266	-0.469	-0.012	-0.003	-0.027	0.308	0.095	-0.927	-0.866	0.868

Bold P<0.05
Slope>0
Slope<0

11.3 Nex-NASA based climate indices (1970-2004)

Stations\Index	TMAXmean	TMINmean	su25	id0	tr20	fd0	txx	txn	tnx	tnn	tx10p	tx90p	tn10p	tn90p	wsdi	csdi	dtr	rx1day	rx5day	sdii	r10mm	r20mm	cdd	cwd	r95p	r99p	prcptot
GB & AJK	0.011	0.02	-0.002	-0.112	0	-0.075	0.01	0.009	0.023	0.012	-0.083	0.245	-0.209	0.365	0.388	-0.139	-0.009	0.042	0.188	0.003	0.013	0.004	-0.016	0.101	0.175	-0.257	1.023
Khyber Pakhtun	0.011	0.018	0.302	0	0.112	-0.418	0.005	0.01	0.009	0.013	-0.13	0.183	-0.224	0.329	0.447	-0.32	-0.007	0.046	-0.03	0.004	0.034	0.006	-0.045	0.227	0.789	0.317	1.334
Punjab	0.013	0.018	0.256	0	0.089	0	0.008	0.014	0.011	0.02	-0.194	0.186	-0.251	0.339	0.258	-0.492	-0.005	0.021	-0.046	-0.003	-0.008	-0.004	-0.073	0.048	-0.365	0.033	0.03
Sindh	0.017	0.02	0.265	0	0.153	0	0.009	0.009	0.01	0.028	-0.25	0.327	-0.391	0.396	0.381	-0.336	-0.003	0.113	0.183	0.004	0.004	0.007	1.23	-0.097	0.101	0.223	-0.409
Baluchistan	0.014	0.021	0.177	0	0.29	0	0.006	-0.015	0.014	0.019	-0.139	0.332	-0.288	0.424	0.461	-0.304	-0.007	0.032	0.038	0.003	0.002	0	-0.42	0.018	-0.002	0.091	-0.091

Bold P<0.05
Slope>0
Slope<0

11.4 CORDEX based climate indices (1970-2004)

Stations\Index	TMAXmean	TMINmean	su25	id0	tr20	fd0	txx	txn	tnx	tnn	tx10p	tx90p	tn10p	tn90p	wsdi	csdi	dtr	rx1day	rx5day	sdii	r10mm	r20mm	cdd	cwd	r95p	r99p	prcptot
GB & AJK	0.024	0.026	0.221	0	0.251	-0.228	0.012	0.047	0.02	0.028	-0.148	0.173	-0.267	0.207	0.485	-0.483	-0.002	0.05	0.627	0.017	0.132	0.031	0.039	-0.076	1.837	0.936	1.927
Khyber Pakhtun	0.028	0.025	0.352	0	0.71	-0.36	0.015	0.063	0.028	0.022	-0.272	0.354	-0.209	0.258	0.448	-0.204	-0.003	-0.013	0.087	0.004	0.008	0	0.041	-0.169	0.32	0.268	0.371
Punjab	0.027	0.028	0.473	0	0.318	-0.027	0.03	0.058	0.008	0.035	-0.283	0.296	-0.301	0.459	0.369	-0.282	-0.001	0.031	0.103	0.005	0.005	0	-0.131	-0.049	0.675	0.25	1.065
Sindh	0.027	0.028	0.527	0	0.417	0	0.033	0.048	0.014	0.051	-0.338	0.354	-0.404	0.581	0.517	-0.534	-0.001	0.014	0.811	0.023	0.008	-0.005	0.807	0.018	0.69	0.794	-0.333
Baluchistan	0.031	0.029	0.394	0	0.425	0.014	0.02	0.042	0.017	0.029	-0.35	0.376	-0.389	0.526	0.585	-0.622	0.002	0.024	0.064	0.003	0	0	0.209	0.009	0.118	0.036	0.092

Bold P<0.05
Slope>0
Slope<0

11.5 Nex-NASA Based Projected Climate Indices (2016-2045)

RCP 4.5

Stations\Index	TMAXmean	TMINmean	su25	id0	tr20	fd0	gsl	txx	txn	tnx	tnn	tx10p	tx90p	tn10p	tn90p	wsdi	csdi	dtr	rx1day	rx5day	sdii	r10mm	r20mm	R25mm	cdd	cwd	r95p	r99p	prcptot
GB & AJK	0.035	0.034	0.034	-0.287	0	-0.582	0.555	0.037	0.026	0.033	0.05	-0.468	0.425	-0.54	0.513	0.51	-0.662	0.001	-0.179	-0.187	-0.002	-0.062	-0.026	-0.018	-0.176	0.192	-1.342	-1.117	0.087
Khyber Pakhtun	0.039	0.038	0.589	0	0.959	-0.568	0.804	0.03	0.053	0.045	0.059	-0.484	0.533	-0.556	0.628	0.705	-0.796	0	-0.102	0.178	0.007	0.007	-0.002	-0.001	-0.106	0.021	-0.171	-0.375	0.747
Punjab	0.032	0.036	0.282	0	0.407	0	-0.002	0.017	0.059	0.026	0.049	-0.442	0.504	-0.59	0.615	0.72	-1.007	-0.004	-0.001	0.098	0.012	0.069	0.003	0.001	-0.228	0.357	0.838	0.338	1.857
Sindh	0.027	0.053	0.522	0	0.327	0	-0.002	0.043	0.042	0.011	0.032	-0.37	0.564	-0.696	0.669	0.708	-0.684	-0.006	0.126	0.629	0.015	0.028	0.002	0.002	-0.252	0.16	0.538	-0.089	2.105
Baluchistan	0.034	0.04	0.394	0	0.796	0	-0.002	0.024	0.056	0.031	0.045	-0.397	0.613	-0.644	0.689	0.808	-0.765	-0.006	0.034	0.25	0.005	0.004	0	-0.002	-0.398	0.063	0.143	0.16	0.669

RCP 8.5

Stations\Index	TMAXmean	TMINmean	su25	id0	tr20	fd0	gsl	txx	txn	tnx	tnn	tx10p	tx90p	tn10p	tn90p	wsdi	csdi	dtr	rx1day	rx5day	sdii	r10mm	r20mm	R25mm	cdd	cwd	r95p	r99p	prcptot
GB & AJK	0.045	0.054	0.172	-0.718	0	-0.905	0.896	0.048	0.061	0.041	0.077	-0.563	0.608	-0.644	9.788	1.082	-0.619	-0.009	0.091	0.029	0.004	0.093	0.035	0.005	-0.092	0.081	1.575	0.789	2.675
Khyber Pakhtunkhwa	0.044	0.051	0.407	0	1.098	-0.731	0.664	0.052	0.044	0.037	0.065	-0.492	0.537	-0.662	0.796	0.869	-0.916	-0.007	0.01	-0.05	-0.004	0.055	-0.01	-0.001	-0.112	0.27	0.139	-0.046	1.415
Punjab	0.037	0.047	0.677	0	0.3	0	-0.002	0.032	0.029	0.055	0.043	-0.419	0.586	-0.653	0.829	0.99	-0.792	-0.009	-0.126	-0.538	0	-0.082	-0.024	-0.005	-0.638	0.286	-0.713	-1.488	0.402
Sindh	0.033	0.059	0.467	0	0.62	0	-0.002	0.021	0.015	0.091	0.005	-0.406	0.626	-0.625	0.827	0.657	-0.451	-0.006	0.046	-0.071	0.009	-0.012	0.006	0	1.786	0.043	0.016	0.257	0.487
Baluchistan	0.038	0.043	0.872	0	0.527	0	-0.002	0.046	-0.003	0.039	0.006	-0.408	0.626	-0.622	0.804	0.685	-0.621	-0.005	0.007	-0.016	-0.002	0	0	0	0.574	-0.026	-0.034	-0.025	-0.286

Bold: P<0.05; Slope>0 (orange); Slope<0 (blue)